Death Rays, Jet Packs, Stunts & Supercars

Death Rays, Jet Packs, Stunts & Supercars

THE FANTASTIC PHYSICS OF FILM'S MOST CELEBRATED SECRET AGENT

Barry Parker

THE JOHNS HOPKINS UNIVERSITY PRESS

Baltimore

© 2005 The Johns Hopkins University Press
All rights reserved. Published 2005
Printed in the United States of America on acid-free paper
9 8 7 6 5 4 3 2 1

The Johns Hopkins University Press
2715 North Charles Street
Baltimore, Maryland 21218-4363
www.press.jhu.edu

Library of Congress Cataloging-in-Publication Data

Parker, Barry R.
Death rays, jet packs, stunts, and supercars : the fantastic physics of film's most celebrated secret agent / Barry Parker.
    p.   cm.
Includes bibliographical references and index.
ISBN 0-8018-8248-6 (hardcover : alk. paper)
1. Technology. 2. Science. 3. James Bond films. 4. Bond, James (Fictitious character). I. Title.
T49.5.P374 2005
600—dc22   2005007782

A catalog record for this book is available from the British Library.

Illustrations are by Lori Scoffield Beer

Contents

Acknowledgments

I am grateful to Trevor Lipscombe for his many suggestions and help in preparing this volume. I would also like to thank Nancy Wachter for careful editing of the manuscript, and the staff of the Johns Hopkins University Press for their assistance in bringing this project to completion. I would also like to thank my artist, Lori Beer, for doing an excellent job on the drawings. For more information on James Bond and information on other books by the author, visit the web page www.BarryParkerbooks.com.

Introduction

The first film I saw featuring James Bond was *From Russia with Love* in 1963, and it's still one of my favorites. I can't say I became a Bond fan at that time because I had no idea Bond movies would continue for as long as they have. Seeing *Goldfinger*, however, I was hooked. To prepare for this book I watched the twenty Bond movies several times and read all the Bond novels by Ian Fleming, who originated the character in 1953.

My interest here is the science and technology—or more precisely, the physics—in the Bond movies, which no doubt helped make them popular. As you might expect, some of it was good, valid science, and some was a little far-fetched. I wasn't too perturbed by the bogus science, because even when the science was bad, the movie itself was usually pretty good. Nevertheless, I think it's important to point out what the films got right and what they didn't. I will also provide some background on the science. It's always interesting to find out how things work and the basic principles involved.

British writer Ian Fleming named his hero after the author of a book on birds titled *Birds of the West Indies,* by ornithologist James Bond. Fleming later met the "real" James Bond, who was honored that his name had become so famous.

CHAPTER

Bond for Beginners

hen Albert Broccoli and Harry Saltzman decided in 1961 to make a movie of Ian Fleming's novel *Dr. No*, they had no idea how successful it would be. Not only have most of Fleming's novels been made into movies, but James Bond is still as popular as ever. What is even more amazing is that the movies have been tremendously successful despite their similarities. In other words, they're "formula" movies: in each of them, James Bond is sent on assignment, and after almost getting killed several times he ends up saving the world, or at least a city or two (table 1). Along the way, he uses all kinds of high-tech gadgets and fancy cars, and, in the end, he always gets the girl. To begin our adventure with the fantastic physics of the incomparable James Bond, I give a brief overview of each movie's plot and the physics in the film.

The Adventure Begins: *Dr. No, From Russia with Love,* and *Goldfinger*

In *Dr. No* Bond is sent to Jamaica to investigate the death of John Strangways, a secret service agent, and his secretary. He is also asked to look into the disappearance of several rockets that the United

Table 1 Movie title, year, and actor who played James Bond

Movie Title	Year	Bond Actor
Dr. No	1962	Sean Connery
From Russia with Love	1963	Sean Connery
Goldfinger	1964	Sean Connery
Thunderball	1965	Sean Connery
You Only Live Twice	1967	Sean Connery
On Her Majesty's Secret Service	1969	George Lazenby
Diamonds Are Forever	1971	Sean Connery
Live and Let Die	1973	Roger Moore
The Man with the Golden Gun	1974	Roger Moore
The Spy Who Loved Me	1977	Roger Moore
Moonraker	1979	Roger Moore
For Your Eyes Only	1981	Roger Moore
Octopussy	1983	Roger Moore
A View to a Kill	1985	Roger Moore
The Living Daylights	1987	Timothy Dalton
Licence to Kill	1989	Timothy Dalton
GoldenEye	1995	Pierce Brosnan
Tomorrow Never Dies	1997	Pierce Brosnan
The World Is Not Enough	1999	Pierce Brosnan
Die Another Day	2002	Pierce Brosnan

States has sent over the Caribbean as part of its space program. Attempts are made on Bond's life almost from the moment he steps from his plane in Jamaica, and he discovers that a mysterious Dr. No may be involved. Bond learns that Strangways died soon after visiting Crab Key, an island Dr. No owns near Jamaica. When he checks into the boat Strangways hired, he discovers ore samples from the island have left behind traces of radioactivity.

Bond decides to visit Crab Key and hires Quarrel, a fisherman who is actually a CIA agent, to take him there. They arrive late at night in an effort to avoid detection. The next morning they encounter a beautiful young girl gathering shells on the beach. Her name is Honey

Ryder, and she tells them that a flame-breathing "dragon" chases people from the island and warns them that they are in danger. Bond, Quarrel, and Honey are detected by a gunboat and fired upon, but they manage to escape. Later the "dragon" arrives and kills Quarrel with its flame. Bond and Honey put up a good fight but are finally captured and taken to Dr. No's lair at the far end of the island. Within the lair is an elaborate laboratory and control room that contains a nuclear reactor.

Bond discovers that Dr. No is the one interfering with the American rocket program. Locked in a small cell, he manages to escape through the vents and makes his way to the control room where Dr. No and his henchmen are busy intercepting another American rocket. Bond throws a switch that creates dangerous radiation levels and everyone except Dr. No rushes for cover. Bond and Dr. No fight on a small platform that is slowly sinking into the boiling water of the reactor. Bond narrowly escapes, but Dr. No disappears into the bubbling water. In the last scene the lab blows up, but Bond and Honey escape in a small boat.

Dr. No was an excellent start and turned out to be more successful than Broccoli and Saltzman had hoped. Indeed, it was the hit movie of 1962, and as a result they were soon busy on a second Bond movie. Much of the success of the film was no doubt due to the actor they chose to play Bond, Sean Connery. Although he was almost an unknown at the time, he was ideal for the role; to many people it soon seemed that Connery *was* Bond.

The second Bond film was *From Russia with Love*. Interestingly, *Dr. No* was not the first Bond novel that Fleming wrote but the sixth, and *From Russia with Love* was the fifth. This was not a problem, however, because there was no direct link between the stories as there was between some of the novels. In *From Russia with Love*, Bond is lured to Istanbul, Turkey. The secret service in London has learned that a woman in the Russian embassy in Istanbul has seen Bond's photo in a file and claims she has fallen in love with him. She wants to defect and will bring the Russian encryptic machine known as Lektor with her if Bond will come to Istanbul and take her to London. M, Bond's boss, calls him and tells him about the offer. Bond is sure it is a trap, but when he sees a picture of the Russian girl he decides to take the assignment anyway.

It is a trap, but strangely, it isn't the Russians who want to kill Bond. Instead, an organization made up of Russian mafia and other criminals called SPECTRE wants him dead, and they convince an attractive embassy worker, Tatiana Romanova, to join their scheme. As it turns out, she does not know they are associated with SPECTRE.

Bond meets the Turkish agent Kerim Bey in Istanbul and they devise a plan to get the Lektor. The plan is successful. Bond and Tatiana flee Turkey on the *Orient Express*, but both Russian and SPECTRE agents are also on board and the action soon picks up. One of the best fight scenes in the entire series takes place on the train when Bond encounters Red Grant, an agent from SPECTRE. Bond manages to kill him, but Bond's troubles are not over. He and Tatiana are attacked by a helicopter, then by a fleet of boats, but they manage to get to London. Even in London they are attacked by another SPECTRE agent, Rosa Klebb, who has a poison-tipped blade hidden in her shoe.

Although *From Russia with Love* is one of the better Bond movies, it does not have a lot of physics in it. Bond is given a briefcase by the Secret Service's equipment officer, Major Boothroyd—or Q as he is better known—that contains several interesting gadgets, and the Lektor no doubt involves some physics, but that's about it.

These first two Bond movies were so successful that a lot more money was poured into the third in the series. *Goldfinger*, named after the villain of the film, is still considered to be one of the best. It begins with officials of the Bank of England discovering that someone is stockpiling vast quantities of gold, and they suspect Auric Goldfinger. They also believe he is smuggling gold out of the country, but they don't know how he is doing it. Bond is sent to investigate. (Actually, Bond has already had an encounter with Goldfinger in Miami, where he puts a stop to his cheating at cards.) He meets Goldfinger at a country club, where he plays golf with him, and again Goldfinger tries to cheat, but Bond outcheats him and wins. Goldfinger is quite annoyed by now. After the golf match, Bond puts a tracking device in Goldfinger's car and tracks him to his factory in Switzerland.

Goldfinger realizes he has been trailed and sends his men after Bond. But Q has supplied Bond with an Aston Martin DB5 that has an amazing array of weapons, including a machine gun, an ejector seat, a tire slasher, and a bulletproof shield, and Bond makes good use of them. Even with these gadgets, Bond is finally captured, but this takes us to one of the best "science scenes" in the series. Goldfinger decides to use a laser to kill Bond. At the time, lasers were new and the one Goldfinger used appears to be quite menacing. But with some quick thinking Bond manages to avoid it. Just as Goldfinger is walking away from him, Bond mentions "Operation Grand Slam," which is a secret operation that Goldfinger has been planning for some time. Goldfinger is quite surprised to hear that Bond knows about it (actually, Bond has only overheard the name).

Operation Grand Slam is a planned raid on Fort Knox in the United States. Goldfinger intends to explode an atomic bomb at Fort Knox, which he believes will increase the value of his stockpile of gold. Assisting him in the enterprise is an attractive associate named Pussy Galore, who is the leader of a flying circus. Pussy and her pilots are to fly over Fort Knox and release nerve gas.

The raid appears to be a success at first. The guards all appear to be unconscious, and Goldfinger blasts his way into the gold depository by using a laser. But after a "tussle" in the hay with Bond, Pussy had decided to change sides. So while Goldfinger and his team are setting up the atomic bomb, they are attacked by U.S. soldiers (who had been faking unconsciousness). Bond, of course, also takes part in the fight. In fact, in one of the critical scenes, he fights Goldfinger's much-feared assistant, Oddjob, and although Bond is no match for him physically, he manages to electrocute him. Goldfinger escapes, but in the last scene Bond meets up with him again. Bond is being flown to Washington, and to his surprise Goldfinger is on the plane. After a brief fight, however,

Goldfinger comes to a fitting end: he is knocked out of the plane without a parachute.

Physics plays an important role in the movie. The laser used by Goldfinger on Bond had just been invented at the time, and there was a lot of interest in it. Goldfinger also used a laser to get into Fort Knox.

Amazed by Success: *Thunderball, You Only Live Twice,* and *On Her Majesty's Secret Service*

Bondmania was now at its height. *Goldfinger* was a tremendous success, and Bond fans could not wait for the next Bond movie to appear. And it soon came. The next movie was even more elaborate and expensive to produce, but earlier ones had made so much money, it hardly seemed to matter.

The next in the series, *Thunderball,* started with one of the most exciting stunts that the series had seen. After a fight in which he kills a man, Bond seems to be cornered, but he straps something on his back, which turns out to be a "jet pack," and he blasts off over the fence to his car—a spectacular beginning.

In *Thunderball,* SPECTRE agents hijack a British Vulcan bomber carrying two nuclear bombs. They switch the pilot and he, in turn, kills the others in the plane, flies the plane to the Bahamas, and crash lands it in shallow water. Largo, the villain, is waiting for him on his yacht, the *Disco Volante.* He plans on using the nuclear bombs to threaten Miami. Bond is sent to investigate and becomes suspicious of Largo. He meets Largo's girlfriend, Domino, while she is scuba diving (her foot gets stuck in some coral and he frees it), and that evening he encounters her again with Largo in a local nightclub. The next day he visits Largo at his villa at Palmyra. That night he checks out the *Disco Volante* using scuba gear. He finds that Largo and his crew of frogmen are soon to leave on a mysterious mission.

Bond informs his friend Felix Leiter of the CIA about Largo's activities. Using a helicopter, they locate the hidden bomber with camouflage netting over it, so they have a good idea where Largo is headed. Leiter then organizes a team of frogmen. Some of the most exciting scenes of the film take place when Leiter's team encounters Largo's men. An exciting underwater fight ensues, with Bond taking out several of Largo's frogmen. When he sees the fight going against him, Largo escapes by releasing the rear end of the *Disco Volante* (the front section is a speedy hydrofoil). Bond, however, manages to cling to the hydrofoils of the fleeing boat (quite a feat, when you think about it) and eventually climbs aboard. He fights with Largo, but Largo gets the drop on him and is ready to shoot him when suddenly Domino appears. She shoots Largo with a harpoon gun. At this point the hydrofoil is headed for a reef and it is too late to do anything but abandon it, so Bond and Domino jump overboard—just in time.

The movie contains a lot of interesting physics, starting with the jet pack. Largo's hydrofoil involves several important physical principles, as do the various underwater devices that are used.

The next Bond movie, *You Only Live Twice*, takes place in Japan. It opens with a manned American rocket being "gobbled up" by a mysterious spacecraft. The mysterious spacecraft then heads back to Earth. The Americans suspect the Russians, but later the Russians also lose a spacecraft in the same way, and they blame the Americans. It looks like World War III could begin soon, but the British are suspicious. One of their observers believes he saw one of the rockets disappear into the Sea of Japan.

Bond, as it turns out, is in Hong Kong, and he is sent to investigate. He teams up with the Japanese agent Tiger Tanaka. The trail eventually leads to an island off the coast of Japan, and Bond uses what looks like a miniature helicopter called "Little Nellie" to fly reconnaissance over the island. Four helicopters attack him, but, like his cars, his helicopter also has an amazing array of weapons (a machine gun, a heat-seeking missile, and a flamethrower) and with them he manages to down all four helicopters.

Bond becomes even more suspicious of the island. To get a

closer look at it without causing a lot of suspicion he goes to a nearby village, disguises himself as a Japanese fisherman, and marries a local Japanese girl, Kissy Suzuki. Along with Kissy, Bond goes to the island and climbs the volcano near its center. In the calderas of the volcano is what appears to be a lake, but they discover it is actually a blue sliding platform. They see a helicopter disappear into it. Bond sends Kissy to get Tiger and his Ninja fighters, and he breaks into the volcano. Inside he finds a large control room and rocket-launching pad. He knocks out one of the workers and takes his uniform, but Blofeld, who is at the control console, soon notices Bond and summons him.

Blofeld and his crew are about to send a rocket to capture another spacecraft, and Blofeld invites Bond to watch. It seems that Blofeld knows who Bond is at this point, and, indeed, he greets him with, "So, we meet at last." After what Bond has done to Blofeld in the past few movies, this seems pretty tame. Bond watches for a while, then suddenly jumps forward and throws the lever to open the top of the volcano. Blofeld is caught off guard but recovers and closes it. A few of Tiger's Ninja fighters get in but they are quickly overpowered by Blofeld's men. One of them, however, manages to place an explosive charge on the roof and blows a hole in it; dozens of ropes suddenly appear, and Ninjas come streaming down. This is one of the most amazing scenes in the movie. A battle begins, and Blofeld's men are overcome.

Nevertheless, Blofeld has a few more tricks up his sleeve. He orders his men to seal the control room with steel shutters and he escapes. Bond is now outside the control room, but finally manages to get into it through a secret entrance, and with only seconds to go he stops the SPECTRE spaceship from capturing the American space capsule. In the final scene the volcano blows up, and Bond, Kissy, Tiger, and the Ninjas escape.

The physics in this movie centers on the space scenes and the rockets, but the helicopter "Little Nellie" is also quite fascinating and involves a lot of physical principles.

The Bond series took a new twist in the next movie. Sean

Connery had been saying for some time that he was going to retire from the Bond movies, and this time he apparently meant it. So the search was on for a new Bond. Broccoli and Saltzman finally settled on an Australian, George Lazenby, to be the next Bond. In some respects he was ideal for the role; he was tall, dark, good-looking, and athletic, but he had no acting experience. Furthermore, his Australian accent was a problem. Despite the problems, Broccoli and Saltzman decided to go with Lazenby. The movie he was to star in was *On Her Majesty's Secret Service*; it was one of Fleming's best novels. It was so good, in fact, that unlike previous Bond movies, they did not deviate from the novel.

With such a good story line, everything should have gone smoothly, and the movie should have been a tremendous success. But the shock brought about by Sean Connery's absence, coupled with Lazenby's inability to act, did not allow this to happen. The movie was not a failure, but it wasn't the hit the previous Bond movies had been. Over time, its stature has increased, however, and it is now a favorite of many Bond fans.

On Her Majesty's Secret Service begins with Bond rescuing a damsel in distress. She appears to be trying to drown herself, but after Bond rescues her, they are attacked by thugs. Bond fights them off, but while he is busy with them, the girl drives off in his car. The girl turns out to be Tracy Di Vicenzo, the daughter of Marc Ange Draco, the head of a powerful crime syndicate. Bond encounters Tracy again a little later in a gambling casino in Portugal, and he also meets her father. Draco is impressed with Bond and offers him a million pounds to marry his daughter. Bond turns him down, but realizes that Draco may have information on the whereabouts of Blofeld and, as it turns out, he does. Bond finds out that Blofeld is at a resort in Switzerland (fig. 1) and that he is claiming that he is rightfully a count and is having his claim investigated by the College of Arms in London. Bond decides to pose as Sir Hilary Bray, the investigator of the claim.

Bond goes to Switzerland and meets Blofeld's assistant, Irma Bunt; she takes him in a helicopter to Piz Gloria, Blofeld's new head-

Fig. 1. Piz Gloria, Ernst Stavro Blofeld's hideout in the Swiss Alps in *On Her Majesty's Secret Service*

---

quarters. Piz Gloria is an awe-inspiring site, perched on a snow-covered peak in the Alps. Bond meets Blofeld, but strangely, they don't appear to know one another, despite the fact that they met in the last movie. Bond is surprised to find a large group of beautiful girls at Piz Gloria. He is told that it is an allergy clinic, and the girls are all patients. With so many beautiful girls around it doesn't take Bond long to form a liaison with one of them, and he is caught in her room. Blofeld soon realizes who Bond really is and has him locked up, but, as usual, Bond escapes and the rest of the movie is filled with action. Bond steals some skis and takes off down the mountain with Blofeld and his men in hot pursuit. The following ski scenes are extraordinary; they contain some of the best ski scenes in the series. Bond narrowly escapes his pursuers and finally makes his way to the village below.

The chase continues in the village, where Bond meets up with

Tracy (what a coincidence!) at a skating rink. Blofeld and his men now surround them, but Tracy has a car and they make a run for it. An exciting car chase ensues, but they eventually crash. Then, both Tracy and Bond are chased on skis, and Blofeld triggers an avalanche in an effort to stop them. Tracy is captured, but Bond gets away. Bond then goes to Tracy's father who organizes a helicopter attack on Piz Gloria. An exciting fight scene follows with Blofeld escaping on—of all things—a bobsled.

The physics in the movie is associated with the stunts, in particular, those involving skiing. Few technical gadgets are used, but Bond does use one for breaking into a safe.

Worries about a New Bond: *Diamonds Are Forever, Live and Let Die,* and *The Man with the Golden Gun*

As I mentioned, *On Her Majesty's Secret Service* did not do very well at the box office, and the basic problem seemed to be Lazenby. Lazenby himself decided that this would be his one and only Bond film. A search for a new Bond was initiated, but it hadn't gone far when Broccoli and Saltzman decided to try once more to get Connery back. They made him an offer he couldn't refuse, and it worked; he said he would come back for one final movie. It wasn't as good as his previous ones, but that didn't seem to matter; everyone was happy just to have him back. The movie was *Diamonds Are Forever.*

The film begins with members of the British government discovering that large shipments of diamonds are disappearing. They believe that a smuggling ring is operating and send Bond to investigate. As it turns out, there is much more to it than a smuggling ring. Bond's investigation takes him to Las Vegas, and much of the movie centers around Las Vegas and its glittering lights. The trail leads to Willard Whyte, a billionaire recluse who resides in the penthouse of one of the Las Vegas towers. Bond scales the outside of the building and breaks into Whyte's apartment and, to his surprise, he encounters not Whyte but Blofeld. Blofeld has taken Whyte's place and imprisoned him in a desert hideaway.

Bond learns that the diamonds have been incorporated into a satellite laser. This laser is so powerful it can bring down missiles and planes, and Blofeld later proudly demonstrates it to Bond. Bond learns that Blofeld is planning to use the laser to decimate Washington, D.C., unless his ransom demands are met. Bond flies into action, shooting Blofeld in the head, only to discover that he has a double. The real Blofeld forces Bond into an elevator in which he is gassed; he is then taken to the desert where he is buried alive. Again Bond escapes. Blofeld has now disappeared, but Bond finds out that Willard Whyte is imprisoned at his summer home on the desert. He visits Whyte and learns that Blofeld is likely on an oil rig off the California coast. Bond flies a helicopter to the rig. The last scenes involve a fight on the oil rig and end with the rig being destroyed, but the ever-elusive Blofeld escapes again. A lot of the physics in the film centers on the laser in the satellite, but some of the interesting devices used also involve physics.

Connery had now decided that his retirement would be final, so a search for a new Bond took place. The man Broccoli and Saltzman settled on was Roger Moore. Unlike Lazenby, Moore had considerable acting experience, and he was relatively well known. His style was quite different from Connery's, though. Light comedy was his specialty, and even when he tried to act tough, it was hard to take him seriously. He just didn't have the "tough-guy persona" that Connery did. As a result, the emphasis in the next Bond movies changed considerably.

Connery had used some humor—mostly one-liners—but humor became a staple of Moore's movies and because of this it was difficult to take his movies seriously. Oh, I know . . . we're not supposed to take Bond movies seriously, but even so, with all the humor and slapstick it was hard to make them suspenseful. Moore was in some of the better Bond movies, but he was also associated with the worst ones. His first effort, *Live and Let Die*, in general, was considered to be reasonably successful. In it British agents investigating a drug-smuggling operation in the United States are killed, and the killings appear to be linked to a Harlem boss referred to as Mr. Big (he is also known as Dr. Kananga). Bond is sent to investigate. After several adventures in Harlem, Bond is captured and taken to Kananga's island, San Monique. In one scene on San Monique he is trapped on a small island surrounded by alligators, but he manages to escape by jumping across their backs.

The film also stars Jane Seymour as Kananga's fortune-teller, Solitaire. Kananga is flooding the United States with heroin and relies on Solitaire's fortune-telling to keep him ahead of the law. Kananga's hideaway is an interesting place. It is a large cave with a graveyard at the entrance, and it is here that Bond fights Kananga to the death.

Live and Let Die was reasonably well received at the box office. It may not have a lot of physics in it, but there are some exciting boat chases through the bayous that involve the principles of physics.

Moore's second Bond movie was *The Man with the Golden Gun*. In the film a gold bullet arrives at the Secret Service office in London with "007" carved on it. The British believe that it is a message telling them that Bond is the next target of the assassin Scaramanga, but no one has any idea what Scaramanga looks like or where he resides. Bond is sure, however, that he can find him, and he starts with the golden bullet. Using it he manages to locate Scaramanga's girlfriend Andrea, and through her he finally locates Scaramanga. His first encounter with Scaramanga comes at a Kung Fu school in Hong Kong. During the next few scenes several attempts are made on Bond's life, one of them during an exciting boat chase. Scaramanga eventually escapes in a car-airplane with female agent Mary Goodnight as his captive, but Bond tracks him to an island near Phuket in Thailand.

Scaramanga has an elaborate solar energy station on the island, which includes a "solar cannon." He has presumably found a method for converting solar energy to electrical energy with almost a 100% efficiency. An interesting feature of Scaramanga's setup is the low-temperature tanks he uses to store energy. Shortly after Bond arrives on the island, Scaramanga uses a solar cannon to destroy his seaplane, so it appears that he has no escape.

Near the end of the film Scaramanga challenges Bond to a duel, where he uses his golden gun and Bond uses his Walther. The match ends in Scaramanga's house of fun, with Bond outsmarting Scaramanga by pretending to be a wax dummy (Scaramanga actually had a wax dummy of him in the house). By then Goodnight has knocked a technician into one of the low-temperature vats, which has caused a runaway reaction. The resulting explosion destroys the island, but Bond and Goodnight escape in Scaramanga's junk.

The movie contains some interesting physics. First, Scaramanga's Solex Agitator, which enables him to convert solar energy efficiently, is of considerable interest. Also, his method of storing the energy is unique. Finally, the 360-degree twist of a car as it flew across a bridge is not only fun to watch, but it also involves some interesting physics.

High-Tech Sets: *The Spy Who Loved Me* and *Moonraker*

With the success of the previous Bond movies, considerable money was available for even more elaborate sets, and they came with the next few movies. The first of these was *The Spy Who Loved Me*, and Moore gives one of his best performances in it. Many of his fans, in fact, consider it to be the best of the Bond movies. It's interesting that, although the title of one of Fleming's books was used, there is no resemblance whatsoever between the novel and the movie.

In the movie a submarine has disappeared. The British fear that someone has perfected a way of tracking submerged submarines and has captured it. The villain, Karl Stromberg, lives in a marine laboratory called *Atlantis*, which resembles a giant black spider (fig. 2). He also has a large supertanker, which he uses for storing his captured submarines.

Submarines are missing from both the British and Russian fleets,

Fig. 2. Karl Stromberg's hideout, *Atlantis*, in *The Spy Who Loved Me*

and the Russians have sent a girl, Anya Amasova to investigate. The British have sent Bond, and Bond and Amasova team up and head to Stromberg's laboratory, which is just off the coast of Sardinia. Bond pretends to be a marine biologist, but Stromberg soon sees through him, and when Bond and Amasova leave, Stromberg orders his men to kill them.

In the next few scenes Bond and Amasova are chased through the mountains by men on motorbikes, then by men in cars, and finally by a helicopter. The car they are driving, a Lotus Esprit, doubles as a submarine, however, and they escape by driving it into the ocean. Like most of Bond's vehicles it is equipped with rockets, and they use one of the rockets to shoot down the helicopter. They return to *Atlantis* under the water but are attacked by frogmen using underwater sledges. The Lotus, however, still has a few more weapons and they use them to escape.

Bond and Amasova then go aboard the USS *Wayne* to search for Stromberg's supertanker. But the supertanker, named *Liparus*, finds them and engulfs them. Bond is soon captured and is forced to watch while Stromberg initiates his plan to start World War III. According to his plan, two of the submarines in the tanker will go to predetermined positions, with one firing a nuclear missile at the United States and the other firing one at the USSR.

Stromberg, with Amasova as his captive, then leaves for *Atlantis*. Bond, however, escapes, and along with the crew of the submarine, he attacks Stromberg's men. An exciting fight follows, with Bond trying desperately to get to the control room to stop the missiles. The control room is heavily fortified, however, and he has to use the high-explosive detonator from one of the nuclear missiles to blow a hole in it. Bond then races to the controls and changes the course of the missiles just in time, redirecting them so that they hit each other's submarine. Finally, he heads for *Atlantis* on a water scooter to rescue Amasova. In the final scenes Bond overcomes Stromberg and his seven-foot henchman Jaws, and he and Amasova escape *Atlantis* just before it is destroyed by a torpedo.

The film contains a considerable amount of science. One of Bond's most interesting devices is the car-submarine "Wet Nellie." And there are, of course, nuclear bombs aboard the submarines.

About this time the extremely successful movie *Star Wars* appeared, and Broccoli decided it would be advantageous to bring out a space-oriented Bond movie. One of Fleming's novels was entitled *Moonraker*. Although it had an excellent story line, it was dated compared with *Star Wars*, so it had to be changed. As it turned out, the basic script was okay, but in the end I thought the movie was ruined with too much slapstick. The set was spectacular with an elaborate space station and several spaceships, and this obviously helped.

The story begins when a Boeing 747 that is carrying a space shuttle crashes into the Atlantic. British divers examine the wreckage and are surprised to find no sign of the space shuttle. Bond is sent to investigate. Suspicion soon falls on Hugo Drax, who manufactures the shuttles. Bond teams up with Holly Goodhead who is on loan to Drax from NASA but is actually a CIA agent. Bond visits Drax and breaks into his safe; clues in the safe lead him to a glass factory in Venice. It is here that a rather far-fetched speedboat chase takes place, along with a lot of slapstick. Bond then goes to Rio, where Goodhead is captured. Again, there is a speedboat chase in which Bond manages to escape by going over a waterfall on a hang glider (where the hang glider suddenly comes from is a mystery).

Bond and Goodhead then smuggle themselves aboard one of the space shuttles and end up at the space station. The space station is spectacular, but as we will see later, some of the scenes on it contain scientific inaccuracies. Bond and Goodhead disguise themselves as astronauts and discover that Drax is planning to destroy all human life on Earth by using a special nerve gas. He is then going to repopulate Earth with a race of "perfect people."

A NASA shuttle arrives on the scene with an army of astronauts and a fight between them and Drax's men ensues. In the final scene Bond must prevent several spheres that hold poisonous gas from reaching the Earth, and he does this by using the laser gun in one of the shuttles.

Despite the difficulties and flaws of the film, it contains a lot of science. The space station, rockets, space shuttles, and lasers all involve physics. Bond also uses many interesting gadgets, but one of the most breathtaking stunts in the Bond films, a skydive, comes at the beginning of the movie, and it involves several principles of physics.

Back to Basics: *For Your Eyes Only, Octopussy,* and *A View to a Kill*

After *Moonraker* the directors decided to get back to something closer to the original Bond films. The result was *For Your Eyes Only,* which is my favorite of the Bond films made by Moore, even though few gadgets were used. The movie begins with a British spy trawler, the *St. Georges,* being sunk off a Warsaw Pact country. On board is a technical encryptic device called ATAC (Automated Targeting Attack Communicator) that the British use for launching their Polaris missiles from their submarine fleet. They are worried that if it falls into the wrong hands it could be used to attack friendly nations. Soviet agents are interested in the device, and Aris Kristatos, a shipping magnate, goes after it, in hopes of selling it to the Russians. Bond is sent to get to it first.

The film contains some fascinating ski scenes (but not as good as those in *On Her Majesty's Secret Service*) and a lot of interesting underwater scenes. The heroine, Melina Havelock, is on a mission of revenge, and unlike many of the Bond women, she is quite believable. This movie's climax takes place in a spectacular setting—a monastery that is located at the top of a steep cliff in Greece (fig. 3). Bond has to climb it to get at the villain, and his climb is, indeed, one of the highlights of the film. At several points he nearly falls. (Moore apparently had a tremendous fear of heights and hated doing these scenes.) In the end, however, he overcomes Kristatos and manages to retrieve the ATAC. In the final scene General Gogol of the KGB arrives in a helicopter to get it, but Bond throws it off a cliff. The physics in the film is associated with the ski stunts, the underwater scenes, the Yellow Submarine, and the ATAC.

With the turnaround, it might be thought that the next Bond film would be even better, but it wasn't. It did contain a lot of action, but it lacked the suspense of *For Your Eyes Only*. It was called *Octopussy*, after the heroine in the film, and much of it takes place in India. At the beginning of the film agent 009 is killed; he dies gripping a Fabergé egg in his hand. The British send Bond to investigate, and Bond soon discovers a connection between the priceless egg and an elaborate smuggling operation. The plot is, in fact, a little overdone, because, at the same time as the smuggling operation has been uncovered, a renegade general is attempting to force U.S. troops out of Europe so that Russia can attack it.

Bond teams up with Octopussy, who owns a circus; in the latter part of the film they are in Germany. The Russian general places a nuclear bomb, which is set to detonate in a U.S. army base, aboard Octopussy's train. The scenes on the train are among the most exciting

Fig. 3. Monastery on a mountain top in Greece in *For Your Eyes Only*

in the film, with a car chase along the tracks and a spectacular fight on the top of the train. In the end, Bond manages to defuse the bomb just in time.

Some interesting physics is associated with an escape near the beginning of the movie. Bond escapes in a minijet, called *Acrostar*. In the final scenes there is an atomic bomb aboard the train that Bond must defuse, and there are the usual gadgets.

The next film was the low point for the Bond series, and it was Moore's last. Most Bond fans consider it to be one of the worst of the series, but it contains a lot of interesting science. It is called *A View to a Kill.*

In the film a silicon chip is captured from the Soviets that is found to be identical with a British chip that is capable of withstanding the intense electromagnetic blast from a nuclear bomb. Max Zorin of Zorin Industries is implicated, and Bond is sent to investigate. Bond finds that Zorin is stockpiling silicon chips, but he is also doing some drilling near the San Andreas Fault in Silicon Valley. In fact, he is planning on setting off the largest earthquake California has ever seen in an attempt to corner the world's silicon chip market. Despite its flaws, the film has some amazing scenery: the last few scenes take place on the top of the Golden Gate Bridge and are quite spectacular.

The film contains lot of interesting science. First, the atomic blast and radiation pulse that is disastrous to electronic devices is basic physics. Bond also uses several interesting devices, and finally there is the San Andreas Fault and the earthquake.

The Serious Bond: *The Living Daylights* and *Licence to Kill*

With Roger Moore's retirement, there was another search for a new Bond. The directors finally settled on the Shakespearian actor Timothy Dalton. He was tall, dark, handsome, and athletic, and, unlike Lazenby, he was an excellent actor. Dalton wanted to play Bond more like Fleming had intended in his novels. This was partly because his experience was in serious movies, and he had some difficulty with comedy. For a lot of the Bond fans, however, the change was a little too dramatic. The next two movies contain few laughs, which was quite a change from Moore's portrayal of Bond.

The first of Dalton's movies was *The Living Daylights*. In it Bond is responsible for the defection of a top Soviet general named Georgi Koskov. Koskov tells him that another general (Leonid Pushkin) has set up a plan to assassinate British agents. Bond is therefore assigned to kill Pushkin, but he becomes suspicious of Koskov and, as it turns out, he is right. Koskov disappears and it is believed he has been kidnapped back by the Russians, but his escape and recapture were actually staged. In reality he is in cahoots with a U.S. arms dealer, Brad Whitaker. They have organized a smuggling operation in which they buy heroin in Afghanistan and sell it abroad at great profit. Bond finds out about it, and together with the heroine (Kara Milovy) and a local chieftain, he attacks the general and his associates, putting an end to their plan.

There is an exciting scene with Bond and Koskov's assistant, Necros, fighting on the netting hanging out the back of a plane. Bond's car is also quite amazing, and he uses several interesting gadgets (key-ring finder, sniper rifle).

Dalton's second Bond movie, *Licence to Kill*, is considered by many to be one of the "darker" Bond movies. It has a lot of action and is a little more violent than most of the earlier Bond movies. And again, it is a very serious film. It begins with the marriage of Bond's friend Leiter. About the same time a South American drug lord, Sanchez, is captured, but he escapes and takes his revenge on Leiter. He kills Leiter's bride and feeds part of Leiter (his legs) to the sharks. Bond is outraged and out for revenge. M tries to stop him, but Bond defects from the British Secret Service and goes on his own. He is assisted by Pam Bouvier, a CIA agent, who turns out to be one of the better "Bond girls," and also, strangely, by Q, who plays a larger role in this movie than he does in most. He supplies Bond with several helpful gadgets.

Bond finds his way into Sanchez's operation by pretending to be interested in joining it. He finds out how Sanchez is smuggling his heroin (he is dissolving it in gasoline, using large tanker trucks to transport it, then precipitating it out when it reaches its destination) and sets out to stop him.

Some of the most exciting scenes are in the latter part of the film when a fleet of Sanchez's tanker trucks race down the hill from his hideout. This sequence contains several breathtaking stunts, and there are some amazing stunts associated with a fight on the back of an airplane.

The Action Increases: *GoldenEye, Tomorrow Never Dies, The World Is Not Enough,* and *Die Another Day*

Dalton retired after his second movie, and he was replaced with Pierce Brosnan, who combined many of the best features of the earlier Bonds. He was handsome and athletic, and he brought some humor back into the films. Over the next few films, however, the action increased dramatically, sometimes at the expense of plot and characterization. The movies became so action-filled that there was hardly time to breathe between scenes. This was in part a reflection of the trend in other movies of the time.

Brosnan's first movie was *GoldenEye*, named for Fleming's resort in Jamaica, where most of the early Bond novels were written. The film has an amazing stunt in the precredit sequence—a bungee jump off a high dam in Siberia—which has to be one of the best stunts ever pulled off in a movie. The story revolves around two Russian military satellites that are capable of massive interference with communications, computers, and other electronic devices on Earth. This is accomplished by an intense electromagnetic pulse.

The GoldenEye discs that control the satellites are stolen by the Russian mafia, an organization that calls itself Janus. Bond is sent to retrieve them and find out what Janus is up to. He is surprised to learn that the head of Janus is a fellow 00 agent that was supposedly killed nine years earlier in Siberia (when Bond took the bungee jump).

Bond goes to Janus' headquarters in Cuba and learns that Janus has threatened to destroy all the electronic devices in London unless he is paid a ransom. The movie is filled with action, including a downhill car race between Bond and Xenia Onatopp. In the last scenes Bond fights with Janus and members of his organization and destroys him and them.

The film contains a lot of interesting science, starting with the GoldenEye, which completely destroys an outpost in Russia (Severnaya) within seconds. Several interesting gadgets (laser watch, grenade pen) are used, and the exciting bungee jump is at the beginning.

Brosnan's second Bond movie was *Tomorrow Never Dies*. It was another action-packed film with elaborate sets. In it Bond is sent to stop media mogul Elliot Carver from starting a war between the British and the Chinese. Carver has just developed a worldwide satellite news system and wants to control the news of the world.

The movie begins with the sinking of the British warship HMS *Devonshire*. Carver uses his Stealth Ship (fig. 4), a ship that is undetectable to radar and carries a large array of weapons, to drill a hole in the side of HMS *Devonshire*, causing it to sink.

The most spectacular stunt in the movie is called the HALO (high-altitude, low-opening) jump—a sky jump from more than 20,000 feet into the ocean, then down to inspect the HMS *Devonshire*. A Chinese agent Wai Lin is also looking into the incident, and Bond eventually teams up with her. They pursue Carver to Vietnam and, in the end, bring down his empire.

The movie contains a lot of physics. The HALO jump is an excellent stunt that involves physics, and the Stealth Ship avoids radar. Bond's car, a BMW, is also one of the best in the series, and there are a lot of interesting gadgets in the movie (phone with lock-piercing system, a GPS encoder, and a 3D imaging device in the Stealth Ship). And, finally, there is an excellent car chase.

Brosnan's third movie, *The World Is Not Enough*, was like its predecessors in that it contains a lot of action and also involves a lot of physics. It is my favorite of the Brosnan movies. It begins with Sir Robert King of King Industries being killed and Bond being assigned to find his killer. King had been building a huge oil pipeline across Siberia, and his

Fig. 4. Elliot Carver's Stealth Ship in *Tomorrow Never Dies*

daughter Elektra was now taking it over. M and Bond believe that she is in danger and part of Bond's assignment is to protect her.

Soon after Bond meets Elektra, she takes him on skis to inspect the pipeline, but while inspecting the pipeline, they are attacked by para-hawks (snowmachines with parachutes). This is an action-filled sequence, with Bond and Elektra barely escaping their attackers. At one point Bond skis off a cliff onto the parachute of one of the attackers (fig. 5).

The trail leads Bond to an oil-pumping facility in Kazakhstan. Posing as a scientist he meets physicist and nuclear weapons expert Dr. Christmas Jones. Bond also discovers that a known terrorist, and former KGB hitman, Renard, is present at the facility; as it turns out, Renard is

Fig. 5. Bond (Brosnan) jumping off a cliff onto the parachute of a para-hawk in *The World Is Not Enough*

timing. The parachute helped soften Bond's landing, and the fact that he jumped through it helped take one of his attackers out of commission.

Skiing is featured in several of the Bond movies but my favorite ski scenes are in *On Her Majesty's Secret Service*. Bond made his escape from the Swiss chalet Piz Gloria on skis, and the scenes that followed were amazing. He did some fancy skiing, and not all of it on two skis. At one point he lost a ski and had to ski on just one. He didn't make any spectacular jumps, but his skiing and that of his pursuers was fascinating to watch.

If you go to a ski hill today, it is surprising to see that there are almost as many people snowboarding as skiing. One of the Bond movies may have played a role in this. Even before snowboarding was popular it was featured in *A View to a Kill*. This movie wasn't one of my favorites, but the snowboard sequence was great.

The first really amazing stunt executed in the Bond movies was in *Thunderball*. Can anyone forget Bond's remarkable escape with the jet pack at the beginning of the movie? There was a lot of interest in jet packs at the time, and it seemed as if they might be the mobile units of the future. At least, they were fascinating to think about, something the army was doing seriously. They hoped the jet packs might be useful in tight tactical situations on the battlefield, so perhaps it was no surprise that one was featured in a Bond movie.

Anyone who thinks about stunts has to think about skydiving. Jumping out of an airplane at 10,000 feet, rolling around, and doing a few flips is obviously not for the faint-hearted. And being pushed out of an airplane without a parachute would be a nightmare, but that's what happened to Bond in *Moonraker*. Luckily for him, the pilot had jumped out a few seconds earlier with a parachute, and Bond managed to catch up with him and take his parachute. A dive called the HALO jump was featured in *Tomorrow Never Dies*. In this case Bond jumped out of a high-flying airplane over the ocean. He had to avoid radar so he could not open his parachute until he was a few hundred feet above the water. When he hit the water he had to discard his parachute and switch over to a snorkel and tank so he could dive into the ocean to inspect a sunken ship. All in all it was quite a feat. Finally, skydiving is also featured in *GoldenEye* when Bond goes over a cliff on a motorcycle and catches up with a plane and takes it over.

To many, one of the most spectacular stunts of all was the bungee jump in *GoldenEye*. Bond jumps off a 640-foot dam with a bungee cord attached to his body. Some people consider it to be the best stunt ever performed in the movies, and it was, indeed, amazing. I'm not sure which one of the stunts would have been the most difficult to stage, but I do know that all of them involve physics, and that's what we'll be concerned with in this chapter.

Swoosh: Skiing and Ski Jumping

As an avid skier, I enjoyed the skiing scenes in the Bond movies. After watching one of them, I was always anxious to go skiing. Seeing the skiers glide down the mountain so gracefully was a delight. There is a lot of physics involved in skiing, and in ski jumping. A skier accelerates, feels forces on his body, and uses up energy, all of which are important concepts in physics.

Let's start with velocity and acceleration. You are probably familiar with these concepts in relation to cars. A moving car has a velocity; it may, for example, be going 60 miles per hour (mph). To be more precise, we have to specify the direction that it is going; let's say it is going south. When something has both magnitude and direction, as velocity does, we refer to it as a vector, and we frequently use an arrow to designate it. Acceleration is also a vector; it is change in velocity. When you go from 30 mph to 60 mph you have accelerated.

The units of velocity are miles per hour, but we can also express them in other units. Some of these are feet per second (ft/sec), or meters per second (m/sec), or kilometers per second (km/sec) in the metric system.

The relationship between velocity and acceleration is given by

$$v = at,$$

where v is velocity, a is acceleration, and t is time. From this we can determine our velocity at any time if we know our acceleration. It also tells us that acceleration is velocity divided by time, and this means that there is an extra unit of time in acceleration. Acceleration therefore has units of ft/sec/sec or ft/sec^2 (or any of the preceding velocity units with /sec^2 in it).

Let's look a little closer at this formula. Suppose we had an acceleration of 6 ft/sec^2. What would our velocity be at the end of, say, 12 sec? Using the formula we get 72 ft/sec. From this we see that it takes acceleration to get us to a certain velocity, but what causes us to accelerate, or how are we able to accelerate? The answer is that we need a *force*, which you can think of simply as a "push or a pull." In skiing, you need a force to accelerate you down the hill. Where does this force come from? For a skier it is supplied by gravity. Because force gives us an acceleration, and acceleration is a vector, force is also a vector. In other words, it has both magnitude and direction.

The laws governing acceleration and motion, in general, were written down more than three hundred years ago by Sir Isaac Newton. They are now referred to as Newton's three laws of motion.

NEWTON'S FIRST LAW: **A body continues in a state of rest or uniform motion in a straight line, unless acted on by a force.**

This might seem a little strange at first. Does a body in uniform motion truly have no forces acting on it? Indeed, it does not. It took a force to get it up to this velocity (it had to be accelerated), but once the force is released, and no other forces are acting on it, it will remain at this velocity forever. The key words here are "no other forces are acting on it." In practice there will be other forces acting on it; one of the major forces is friction, and as you likely know, friction will eventually stop it. It's pretty hard to imagine any situation where there is no friction, but there are cases where the friction is very low. In the end, though, all objects on Earth are going to stop if no force keeps them going. As we will see later, however, things are a little different in space.

What we would really like to know, though, is what acceleration we get when we apply a force, and this is answered by Newton's second law.

NEWTON'S SECOND LAW: The acceleration produced by a force acting on a body is directly proportional to the magnitude of the force and inversely proportional to the mass of the object.

This might seem like quite a mouthful, and it shocked many people when Newton first announced it. They weren't used to such terms as "directly and inversely proportional to." But these terms really aren't that hard to understand. All "directly proportional to" means is that if we have a quantity X that is directly proportional to Y, then as X increases, Y also increases (for example, if X doubles, so does Y). "Inversely proportional to" means that if X increases, Y decreases.

We can write Newton's second law as force = mass × acceleration, or in algebraic notation as $F = ma$. You must be careful here. The quantity m is mass, which you might think of as weight, but there is a difference. Mass is actually weight divided by the acceleration of gravity (the rate at which objects fall), which is 32 ft/sec^2 on Earth (in the metric system it is 9.8 m/sec^2). What this means is that your mass never varies; it is the same regardless of whether you are standing on Earth, or on a different planet (with a different gravity), or whether you're out in space. Your weight, however, depends on the gravitational field you are in. Gravity varies from planet to planet and is zero in empty space.

NEWTON'S THIRD LAW: For every action there is an equal and opposite reaction.

The action referred to here is the action of a force. Like the first two laws, this one might seem a little strange, but you actually see the results of it almost every day. When you hold a garden hose, you feel a force on your hands as the water gushes out. This is the reactive force, and as we will see later it is this reactive force that makes rockets work.

Because we now know a little about forces, let's look at the forces on a skier as she skis down a mountain. Several forces are involved and, as I mentioned earlier, we can represent them by arrows. They are demonstrated in figure 6.

Gravity, the main force that allows you to ski, acts at the center of gravity of your body; this is where we can consider all the weight of your body (W) to be acting. For convenience we will split W into two forces, F_N and F_S, perpendicular and parallel to the slope. The component parallel to the slope is of particular importance to us; it's the one that gets us down the hill.

Another force that plays a large role in skiing is air resistance. Air resistance has a large effect on how fast you go. It depends on the area of the skier, his velocity, and a constant referred to as the coefficient of drag. Because area is important, a skier can reduce aerodynamic drag by making his effective area smaller. One way of doing this is by going into a tuck position.

Something else that is important in skiing is friction or, more exactly, the frictional force exerted on your skis. It may seem that it should be as low as possible, and during your run downhill you do want it to be low. But at some point you have to stop, and you'll need a lot of friction; this is where the metal edges of your skis come in. The "sharper" they are, the greater the friction, and it's why good skiers keep their skis "tuned." A rapid, parallel turn creates a large amount of friction along the length of the ski.

Friction is determined by what is called the frictional coefficient μ (it can vary between 0 and 1). To make μ very small, racers (and others) apply wax to the bottoms of their skis. This wax repels water and snow at low temperatures and needs to adhere to the skis as long as possible.

In figure 6 we see several other forces. One of them is F_{SNOW}, the force against the front of the ski due to the snow. F_D is the aerodynamic drag force; we will be discussing it in detail later. Finally, if we add these forces together appropriately (subtracting the ones with their arrows in the direction opposite to F_S), we get F_{net}, the resultant force

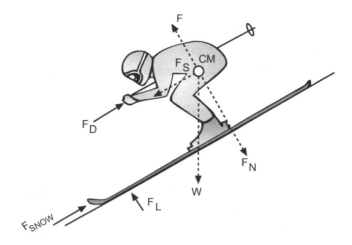

Fig. 6. Forces on a downhill skier. W, weight of the skier; F_N, gravitational force normal to the slope; F_S, gravitational force parallel to the slope; F_D, aerodynamic drag force; F_{SNOW}, snow-plowing force; F, frictional force; F_L, aerodynamic lifting force.

that is acting on the skier. We can then easily determine the skier's acceleration from $a = F_{net}/m$.

Another important concept in relation to skiing is *energy*. Energy comes in several forms, but for us, two forms are of particular importance: *potential energy*, or energy of position (abbreviated as PE), and *kinetic energy*, or energy of motion (KE). Let's begin with potential energy. When we raise a mass m a distance d we do a certain amount of work; this work is defined as the weight of the object (mg) times the distance (d) through which it is raised. When we do work on the object we are increasing its potential energy. It is given by

$$PE = mgh = Wh.$$

In skiing we raise ourselves (and our skis) to the top of the mountain by means of a ski lift, and we gain potential energy. For example, if you weighed 160 pounds and the lift takes you up 1,000 feet, you have a

Fig. 7. Ski jumper

potential energy. As she moves down the run her speed increases, and she converts some energy to kinetic energy. The object, of course, is to jump as far as possible when you hit the end of the run. For this you need as much speed as possible, and this in turn requires as little friction (between the skis and the snow) as possible and a minimum of aerodynamic drag. For maximum lift, it's also important that the skier launch herself forward at exactly the right moment. She does this by getting into a crouch position, with her arms forward and head bent down as if in a dive. As she approaches the end of the run (perhaps 10 feet before the end) she pushes forward with her legs acting like a coiled spring. This gives her the extra lift she needs.

Once in the air, your aim is to stay there as long as possible.

To do this you want to make yourself look as much as possible like an airplane wing, as it has maximum lift. Lean forward and put your skis in the form of a V, with the open end forward (fig. 7). This increases the surface area below your body and provides maximum lift. How far can ski jumpers jump? That depends, of course, on whether they're jumping from a 90-meter or 120-meter ski jump. A good jump for the 120-meter run is 130 meters (429 feet); for the 90-meter run, a corresponding jump would be about 95 meters (313 feet).

Jet Packs

At the beginning of *Thunderball*, Bond is in a church watching a grieving widow leaning over a coffin. The initials on the coffin are JB, which we assume stands for James Bond. What's going on? Is Bond attending his own funeral? Bond follows the grieving widow to a large country mansion. As she enters one of the sitting rooms the camera pans over to a chair, where Bond is already sitting. He gets up and walks over to the widow, saying, "Let me offer my condolences," and then punches her in the face—much to the audience's surprise. The "widow" is actually a man. A fight ensues. Bond eventually overcomes him and breaks his neck, but guards come rushing in. Bond flees across the roof and appears to be trapped, but suddenly he straps on a jet pack and, in a burst of steam, he escapes to his waiting car.

It was certainly a unique way to escape. And there was a lot of interest in jet packs at the time *Thunderball* was being made. The military was particularly interested in them as a possible aid to combat soldiers in tight tactical situations. That was in 1965, and with the tremendous advances in technology since then, you might think that jet packs would now be common. But they're not.

Let's begin with the dynamics of Bond's flight out of the estate. The jet pack applied a force to Bond's body that projected him into an orbit, as shown in figure 8. In physics we refer to this as projectile motion. The math is complicated, but I'm not going to go into much detail.

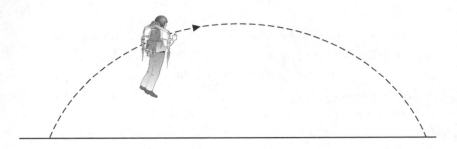

Fig. 8. Track of Bond through the air as he escapes by using a jet pack

Ignoring the matter of air friction, we'll split the initial velocity v into two components: a horizontal one, v_h, and a vertical one, v_v. It's easy to show that the horizontal velocity will remain constant throughout the flight, and the vertical motion will be like that of a falling (or rising) object. If you drop an object from the highest point of the orbit (at the same time the projectile passes), it would strike the ground at the same time the projectile did. My students are always surprised by this when I demonstrate it in class.

Using this information we can determine how far along the horizontal the projectile is at any time and how far it is above the ground. For the first of these we merely multiply v_h by the time it has been in the air. For the vertical velocity at any time we can use $v = v_o - at$, where v_o is the original velocity and a is the acceleration, which in this case is the acceleration of gravity, g, and t is time. Once we have this velocity we can obtain the distance d that the body is above the earth by using the formula

$$d = v^2/2g.$$

For example, assume that we want the maximum height it reaches. If the initial upward velocity is 50 ft/sec, the distance this object will rise above the Earth is $d = 50^2/2(32) = 39.06$ ft.

All these formulas apply to Bond and his flight with the jet pack, but we must be careful. The thrust from the jet pack is applied over a certain distance of the orbit, and these formulas apply only if the blast from the jet is very short, or after the blast is over. And don't forget air friction, which we neglected in our example.

The jet pack that Bond used in *Thunderball* was invented in the 1950s by Wendall Moore of Bell Aerodynamics. Moore himself made some of the first flights with the device, and although he used a safety harness, he hurt his kneecap on one of the tests and was grounded. The tests were continued by one of the other engineers.

The major components of Moore's jet pack were three tanks, which were strapped on the back of the person piloting the device. Two of the tanks contained a 90% solution of hydrogen peroxide, and the third contained nitrogen. At the appropriate time the valve on the nitrogen tank was released, and it forced the hydrogen peroxide into a catalyst chamber, which consisted of fine silver meshes. When the hydrogen peroxide hit the silver it broke down, producing superheated water—or, more exactly, steam at 1,370°F. The steam was forced out through two nozzles that were directed downward under the arms of the pilot (fig. 9). A lot of heat was produced so the pilot had to be protected with a fiberglass shield. With a full load of fuel the device was capable of producing a thrust of approximately 300 lbs for about 20 seconds. This allowed only relatively short flights.

In the Bond film the jet pack was manned by William Suitor, and he still holds most of the records for rocket-belt flight. The device has been improved slightly since 1965, but it is still only capable of short flights. Recently, a similar device, called SoloTrek XFV, was manufactured by Millenium Jet, a California-based company. It is cumbersome, weighs about 350 lbs, and runs on kerosene or other similar fuel. But it can run up to two hours on a tank of fuel. In general, though, it is still in the experimental stages.

NASA is interested in jet packs for the space program. Astronauts are frequently required to work outside of rockets in space and at the space station. They are tethered, but it is well known that if their tether

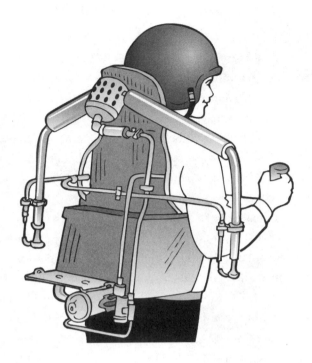

Fig. 9. Details of jet pack. The two jets are shown just below Bond's elbow.

were to break it would be impossible to rescue them. Because of this, a jet pack device called SAFER (Simplified Aid for EVA Rescue) has been developed. It is powered by nitrogen gas thrusters and has 13 minutes of propellant.

The Greatest Thrill: Skydiving

The precredit sequence of *Moonraker* was breathtaking, to say the least. It began with Bond, flying back from an assignment, in an embrace with an attractive woman (as usual). Suddenly, she pulls a gun, and moments later she dons a parachute and jumps out the door. The

Fig. 10. Skydiving. The skydiver holds his hands out to slow his speed.

pilot then appears wearing a parachute, and it is obvious that he is planning to leave too. Bond fights with him briefly, then knocks him out the door. Seconds later Bond is pushed out by the villain Jaws. Unlike the girl and the pilot, however, Bond and Jaws do not have parachutes. It's hard to imagine a worse situation than being a few thousand feet above the earth without a parachute. How does Bond get himself out of this one? From what he does next he obviously knows something about skydiving. He puts his arms to his sides and points his head down and soon catches up with the pilot. After a brief struggle he manages to take the pilot's parachute (fig. 10). A little later Jaws does the same thing and manages to grab onto Bond, but Bond escapes by pulling the ripcord on his chute.

This sequence obviously required some amazing stunt work. Would you like to know how they managed to film the parts where Bond had no parachute? The stuntmen actually had special small parachutes hidden under their jackets. Several takes and a lot of footage were needed before everything came together properly.

So, how did Bond catch up with the pilot? For this the two men had to be falling at different rates. But Galileo showed us years ago that everything falls at the same rate, namely at an acceleration of g. He was, however, neglecting an important factor: aerodynamic drag.

We can easily calculate the speed and distance an object falls if we neglect aerodynamic drag. The formulas on pages 38 and 205 give us the following table.

Time (s)	Velocity (ft/sec)	Distance (ft)
0	0	0
1	32 (22 mph)	16
2	64 (44 mph)	64
3	96 (65 mph)	144
4	128 (87 mph)	256

When the object is dropping through air, however, we must take air drag into consideration, and it can be large. In this case we assumed the acceleration was g (32 ft/sec^2). With air drag it will be given by

$$a = (W - F_D)/m,$$

where W is the weight of the person and F_D is the drag force.

As the skydiver falls faster and faster, the drag force increases, until finally $W = F_D$, and then there is no acceleration. At that point the skydiver has reached what is called *terminal velocity*. This velocity depends on several things, including the cross-sectional area of the skydiver and his velocity. The drag force is given by

$$F_D = c_d \rho v^2 A/2,$$

DEATH RAYS, JET PACKS, STUNTS & SUPERCARS

where ρ is the density of air, c_d is the coefficient of drag, A is the cross-sectional area of the diver, and v is the diver's velocity. Using this we can easily show that his terminal velocity is

$$v_{terminal} = (2W/c_d\rho A)^{1/2}.$$

From this we can see why light objects with large areas, such as feathers and parachutes, reach a terminal velocity so fast. Looking at the formula we see that as A increases and W decreases, $v_{terminal}$ gets smaller. The bigger the area, the greater the air resistance. This resistance exists because the air molecules collide with the falling body, creating an upward force (opposite gravity). Since it depends on the body area of the skydiver, he can slow his speed by going into a "spread eagle"—in other words, by projecting his arms and legs out and maximizing his area in the direction of his fall. Acceleration will decrease and he will reach his terminal velocity relatively rapidly in this position. At any point he can increase his speed by putting his arms to his side and facing his head downward. His new area is just the area of his head and shoulders (much smaller than that of his whole body). Thus, he has considerable control over his speed; he can increase it and lower it at will (within a certain range).

There are two other factors in the preceding formula that I haven't talked about yet, namely, c_d and ρ. The first, c_d, is called the drag coefficient; it depends on the shape of the surface hitting the air, and it ranges from 0 to 1 (actually, slightly more than 1). The drag coefficient is referred to a lot when we talk about cars. For modern cars it is typically .3. There are many ways to control it in skydiving. Special helmets, in the shape of a bullet, are helpful, as are special suits. The second factor, ρ, is the density of air, and because this density decreases as you go upward, away from the Earth, it can be very small at high altitudes. This means that the drag force is much smaller at very high altitudes.

What is the terminal velocity for a typical skydiver? For someone using a spread eagle at moderate altitudes it is on the order of 125 mph. If you pull yourself into a ball so that your area is small, your terminal velocity will be about 200 mph. And if you put your hand at your side and face downward you can travel even faster. This is why Bond was able to catch up with the pilot.

If you go to extreme altitudes, where ρ is very low, you can reach very high terminal velocities. This was done by Joseph Kittinger Jr. in 1960. He made a dive from a helium balloon at 103,000 feet, and it is estimated that he reached a free-fall velocity of 614 mph, which is very close to the speed of sound in air (741 mph). Some people believe that he may have actually broken the sound barrier, but this seems unlikely. He was in a free fall for about four minutes with very little air pressure (at this height ρ is very small). If the air pressure were zero during these four minutes, he would have reached a velocity of more than 5,000 mph, but, of course, it wasn't zero. Kittinger still holds the record for the fastest speed ever achieved by a human in free fall. Plans are underway for breaking his record.

Getting back to the Bond movies, let's look at Bond's HALO (high-altitude, low-opening) jump in *Tomorrow Never Dies*. He had to jump from an airplane at an altitude of 20,000 feet, but he couldn't open his parachute until he was below radar level, which was about 200 feet. His speed would be about 300 mph or more just before his parachute opened. This meant he would hit the water at a bone-shattering speed. At such speeds he obviously had to have a special suit and goggles, and there was still some danger that he might black out for lack of oxygen. Furthermore, when he hit the water he had to get rid of his parachute as quickly as possible, and avoid getting tangled in it. And he had to attach his flippers (they were strapped to his legs) and employ his oxygen tanks. All in all, it was quite a stunt.

Finally, I should mention the fabulous skydive in *GoldenEye* in which Bond goes over a cliff in a motorcycle, then dives down to catch an airplane that is passing. After managing to make his way to the cockpit and climb in, he had to pull the plane out of a steep dive to avoid crashing into the mountain.

Oooh . . . My Stomach: Bungee Jumping

It's hard to say if it was any more death defying, but the stunt at the beginning of *GoldenEye* was certainly the equal of the HALO jump. Bond bungee jumped from the top of a dam in Russia—a drop of 640 feet—down to the bottom of the dam. His speed at the bottom of the free-fall part of the jump had to be near a hundred miles an hour, and this time there wasn't water below—there was cement. The jump was actually performed by stuntman Wayne Michael. A crane was used to prevent Michael from being slammed against the dam. In any bungee jump there are several bounces. To prevent this, Bond used a piton gun; he shot a piton into the cement near the bottom of the dam and it held him until he released himself (fig. 11).

Bungee jumping is a rather recent phenomenon in America. Its origins can be traced to the island of Vanuatu, one of the Pentecostal Islands in the South Pacific. To demonstrate their courage, young men would jump off a high platform with a vine attached to their leg. Their objective was to get as close to the ground as possible without actually hitting it (and possibly killing themselves). The sport was taken up at Oxford in England (by the Oxford University Dangerous Sports Club), and in New Zealand, in the early 1980s. It came to the United States in the early 1990s. Cranes, towers, and hot balloons were used as launching platforms.

What does bungee jumping feel like? According to jumper Michael Black (2001), "You simply have to experience it yourself. . . . The heart-pounding feeling as you leap off the edge is second to none." And since I've never experienced it, I can't elaborate. But this much is known for sure—initially you are in free fall, then the cord begins to stretch. After stretching to its maximum, it pulls you upward, and soon you're going through the same sequence again. You will likely experience three or four of these bounces before you begin to settle down.

Modern bungee cords are made especially for this type of jumping. They are soft and spongy and stretch to three or four times

Fig. 11. Bond as he bungee jumps from a
dam in Siberia

their original length. To understand the physics of bungee jumping we must go back to the concept of energy. We saw that there are two main types of energy (PE and KE), but several other types actually exist. One of these is spring energy (SE), and it is determined by what is called a spring constant k. It is given by

$$SE = \tfrac{1}{2}kd^2$$

where d is the amount the spring (or cord) is pulled from its free or equilibrium length. Let's consider three situations: the jumper ready to jump, the jumper at the end of free fall (at the length of the cord L), and the position at which the cord has stretched to its full amount $(L + d)$ (figs. 12-14). The potential energy $(W(L + d))$ at the end of his fall, when the cord is stretched to its maximum, will be equal to the spring energy of the cord.

$$W(L + d) = \tfrac{1}{2}kd^2$$

This is a relatively simple-looking equation, and it can easily be solved for d. The solution is

$$d = W/k + (W^2/k^2 + 2WL/k)^{1/2}$$

which looks pretty complicated, but it's not too difficult to substitute some numbers. For example, suppose a 180-lb man is jumping from a platform using a 100-foot cord with a k of 5 lb/ft. Inserting this information into the formula above shows that the cord will stretch by 128 feet. (If we apply the formula to Bond's jump we find that a cord slightly longer than 400 feet would be needed.)

With the 100-foot cord, the jumper will be in free fall for the first 100 feet. Neglecting friction, his velocity at the end of this time will be 80 ft/sec or 55.5 mph. Once the bungee cord starts to take up the slack, however, he will quickly decelerate, reaching zero velocity at 228 feet. He will then accelerate upward.

Our calculations are actually quite approximate. In a more detailed calculation, other things would have to be considered. For example, cords of this type frequently have a stiff section near the tower to prevent it from rubbing against the tower; the spring constant of these cords is usually not constant throughout the length of the cord, in many cases it varies; and we neglected air friction in the free-fall section of the jump.

Great Waves: Surfing

The opening of *Die Another Day* had a dramatic surfing sequence (fig. 15). I don't know whether it would classify as a stunt (if the wave was big enough I suppose it would), but it was spectacular nevertheless. The incident was supposed to have taken place along the coast of North Korea, but it was actually filmed in Maui, Hawaii. (I know . . . my reaction was the same . . . Do they actually have waves like that in North Korea?)

Three surfers could be seen in the film. One of them was Laird Hamilton of Hawaii, a giant in the sport of big-wave surfing. He recently surfed the giant waves of Teahupoo, off the coast of Tahiti—something that no one else had ever done. These waves are particularly dangerous because of their volume and speed and because the water is shallow. A fall in a wave there would almost certainly mean death. The waves move so fast that it's impossible to get up on a surfboard. Hamilton solved this by using a jet ski to tow him. It pulled him at the 40 mph needed (the speed of the waves) to get him up on his board. Some people consider this cheating, but all I can say is more power to him.

It might not seem that there is a lot of physics in surfing, but there is. When you watch waves on water, they actually seem to be moving, but a wave is really just a disturbance on the surface of the water. No particles (of water or anything else) move very far from their natural positions. If you placed a cork on the water, it would merely rise and fall as the waves passed over it (there would also be a slight movement back and forth).

Fig. 15. Surfer

Two types of waves exist, *transverse* and *longitudinal*. In transverse waves, the vibrational motion of the individual particles is perpendicular to the direction of travel of the wave. Ocean waves are mainly of this type. The distance between equivalent points on the wave is called its *wavelength* (it is usually designated as λ), and the number of particles passing you per second is called the frequency (f) of the wave. These two things are related to the velocity (v) of the wave by

$$v = \lambda f.$$

In a longitudinal wave the vibration of the individual particles is parallel to the direction of travel of the wave. We represent it as shown in figure 16. The formula above also applies to it. A good example of a wave of this type is sound as it travels through air. This type of wave also occurs, to some extent, on the ocean. In this type of wave there would be some movement of the particles of water in the direction of the wave.

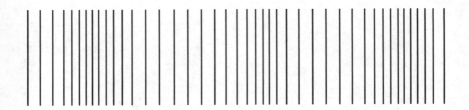

Fig. 16. A longitudinal wave

If you look closely at an ocean wave you see that the particles of water actually move in a circle or an elliptical orbit (fig. 17). In effect, a given particle of water (or cork in the water) moves up and down, and also forward and backward to a small degree. Because they always move back to the original position there is no overall motion of the water.

Although there is no motion of the water, energy does travel through the water. Each particle of the water has energy of vibration and it passes this energy on to its nearest neighbor. It is this energy that allows you to surf.

How do waves arise? They're mostly a result of the wind. How big they become depends on how fast the wind is blowing and how long it blows. Fetch, or the extent of open water over which the wind blows, is also a factor. The larger the open area, the larger the waves can get; this means that in the open sea, waves will get much larger than they will in protected areas. A rule of thumb is that the height of the waves is usually no more than one-half of the wind speed in miles per hour. This means, for example, that a 60-mph wind can generate thirty-foot waves on the ocean.

Surfers love thirty-foot waves. And the real pros hope for fifty

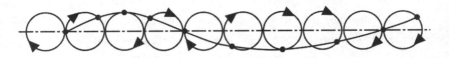

Fig. 17. Movement of water particles in a wave

Fig. 18. A wave breaking on a beach

footers. This leads us to the question: how high can waves get? The highest wave ever recorded was in 1993; it was slightly over 100 feet high, but this doesn't happen very often.

Waves are usually pretty moderate in size when they are in the middle of the ocean. It's when they hit the shallows that they become really big (a wave that generates a tidal wave near shore may only be a few inches high in the middle of the ocean). A wave grows dramatically when its bottom hits the shallows. Exactly when it hits depends on the wavelength of the wave. Waves begin to touch bottom when the depth of the water is half their wavelength. When the bottom of the waves begins hitting, the wave slows down, and other waves soon crowd in behind it, causing the particles of water (which were moving in a circular orbit) to squeeze together and begin moving in an elongated ellipse. The back of the wave is traveling faster than the front at this stage, and this forces it to rise to a peak. But the peak is still traveling fast and it soon curls over and spills forward, creating foam (fig. 18).

Here's a rule that helps us determine when a wave will break. A wave will break when the ratio of the wave height to the water depth is approximately 3:4. In other words, a three-foot wave will break in four feet of water.

Earlier I mentioned that waves carry energy. It is the change in the wave—the sudden increase in size—that occurs when it is near the shore that changes this energy. When the water is deep the energy is mostly potential energy, but when it approaches the shore and breaks, it changes to kinetic energy. At this time the water molecules actually begin to move forward.

The surfer waits for the wave to grow as it approaches land. He wants to catch it as it is cresting, and this is when he gets on top of it. The wave has maximum velocity at this time.

Another principle of physics is involved when the surfer stands up on his surfboard. It is called the Archimedes Principle. This principle states that the weight of the board and the surfer will cause the surfboard to displace an equal weight of water. Because the surfer and his board float, we know that the sum of the forces on the board is zero. From this we can determine

$$\rho_{surfer} = \rho_w - m/dA,$$

where ρ_{surfer} is the density of the surfer, ρ_w is the density of the water, d is the thickness of the board, and A is its area. This is an important formula in the manufacture of surfboards. It tells us that they should be made of very light materials and have a large area. Of course, we don't want the area to be too large, because they would be unwieldy, so surfboards are made of very-low-density fiberglass material, then coated with resin.

So that's it for the stunts. Oh, I know . . . I missed a bunch of them, but I think I covered the more spectacular ones. A few others I should mention briefly are the scenes with Bond clinging to an airplane in both *The Living Daylights* and *Octopussy*. These scenes were spectacular, but they didn't involve a lot of physics.

Death Rays and Ghosts

t was as if some invisible jet impinged upon them and flashed into white flame . . . An almost noiseless and blinding flash of light, and a man fell headlong and lay still; and as the unseen shaft of heat passed over them, pine trees burst into fire . . . " This scene, near the beginning of H. G. Wells's *The War of the Worlds,* introduced us to the "ray gun." And as you might expect, science fiction writers soon jumped on the bandwagon, and everyone, from Flash Gordon to Buck Rogers, was using ray guns. Like time travel, ray guns became a staple of science fiction.

Do ray guns actually exist? We do, in fact, have the key ingredient—a laser beam. Lasers were invented in 1960. Many people were sure we would soon get a ray gun, but, as we will see, there have been problems. Nevertheless, it wasn't long before a laser was used in the Bond movies. *Goldfinger* was released in the early 1960s; and one of the most chilling scenes in the movie was the one in which Bond was almost cut in half by a laser. He was strapped to a table made of gold, and as the villain, Auric Goldfinger, watched, a laser beam moved closer and closer to Bond, melting the table on its way. It was a nice touch by the directors of the film, as lasers had just been invented and there was considerable curiosity about them.

The book on which the movie was based was written in the late 1950s, and Ian Fleming likely knew little about lasers, so as you might expect he didn't use one in it. But what he used was just as deadly and just as chilling. He used a whirling saw blade, which would have cut Bond in half just as fast. For me, the whirling blade was more gruesome. Years ago I visited the site of an accident where someone had fallen onto a large blade while cutting wood, and it was something I hope I never see again. I'm sure I had nightmares about it for weeks.

The whirling saw blade would no doubt have sent chills up the backs of people in the audience had it been used, but the use of the laser in its place was a stroke of genius. The picture of the laser approaching Bond's crotch was used extensively in the advertising and anyone who saw the movie no doubt remembered the scene for a long time. I know I did.

What Is a Laser?

So, what is a laser? Would it really have cut Bond in half? And with the use of a laser in a satellite in *Diamonds Are Forever* and other films, we also have to ask if they are really powerful enough to knock missiles and satellites out of the sky. They can indeed be made powerful, but most are harmless. I used one for demonstrations in class for years, and it always amazed the students when I told them that a laser beam wasn't much different from an ordinary flashlight beam. As long as you didn't shine it in your eye, it wouldn't harm you. In fact, you wouldn't feel anything when it struck your skin.

Let's begin by looking at how laser light differs from ordinary white light. To do this we'll begin with the basic properties of light. In the late 1860s the Scottish physicist James Clerk Maxwell showed that visible light is part of what we now call the electromagnetic spectrum — a spectrum of waves ranging from radio waves to X rays and gamma rays. All these electromagnetic waves, including light, have an associated wavelength and frequency. In figure 19 we see that the wavelength

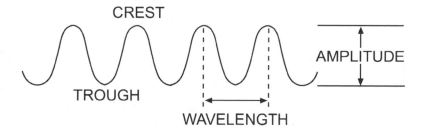

CREST

TROUGH

AMPLITUDE

WAVELENGTH

Fig. 19. A transverse wave showing wavelength and amplitude

of a wave is the distance between equivalent points; for example, it can be the distance between two successive troughs or two successive crests. We can, in fact, identify the type of wave by its wavelength. Radio waves have wavelengths from a mile (kilometer) or more down to a meter or less; visible light, on the other hand, has a wavelength of 10^{-5} cm, and X rays and gamma rays have even shorter wavelengths.

We can also identify electromagnetic waves by their frequency, or number of vibrations per second. All electromagnetic waves travel at the speed of light. In a given time, say a second, a certain number of "crests" would pass us. This number is the frequency. Electromagnetic waves have another property called amplitude; it is associated with the "intensity" of the wave. In the case of ordinary light, the greater the amplitude, the brighter the light. Finally, if we have two waves moving side by side, they may not line up, even if they have the same wavelength. In this case we say there is a phase difference between them.

This is all we need to know about Maxwell's theory for now, but there is a little more about light we need to understand. In 1900 the German physicist Max Planck introduced a theory in which radiation (electromagnetic waves) was considered to be emitted as discrete clumps of energy; he referred to these clumps as "quanta." The energy of these quanta was given by the formula $E = h\upsilon$, where E is energy, υ is frequency, and h is a constant, now called Planck's constant.

Why two different theories? Which one is correct? Actually,

they're both correct. Planck's theory is, in fact, the basis of what is called quantum mechanics, which is now the accepted theory of the world of elementary particles and the interactions between them. Of particular importance, the Danish physicist Niels Bohr used Planck's theory to give us a model of the atom (fig. 20). In this model a "nucleus" of protons and neutrons is surrounded by electrons in various orbits, much like the planets around the sun, but the electron can only orbit at certain fixed distances from the nucleus. Unlike the planets in our solar system, however, the electrons can jump back and forth between the various orbits, but when they do, they emit or absorb a quantum of energy, or "photon," of a particular frequency. More exactly, when an electron in an outer orbit jumps down to a lower orbit it emits, or releases, a photon. When an electron in a lower orbit absorbs a photon it jumps to a higher orbit. These photons, incidentally, are the same as Maxwell's electromagnetic waves.

Bohr's model was an important breakthrough, but it didn't explain everything. In 1916, however, Albert Einstein came to our rescue by putting forward a more complete picture. He asked himself what would happen if a photon struck an electron that was already in an outer orbit (in an "excited state"). The reasonable answer was that it would jump to an even higher orbit. And indeed, this is possible, but Einstein decided that it was also possible that the electron could be forced to a lower orbit, and if so, it would release a photon identical with the one that struck it. We would then have two identical photons.

Einstein called this stimulated emission. When an electron jumps down and releases a photon in the usual way, it is called spontaneous emission. This was a startling idea because it could, in theory, create a "cascade" of photons. How could this happen? Suppose we have a large number of excited atoms (with electrons in outer or excited states). If stimulated emission occurred, each of the released photons would stimulate the release of more photons. In effect, if we started with 2, we would soon have 4, then 8, and so on. The overall effect would be an "amplification" of our beam of photons.

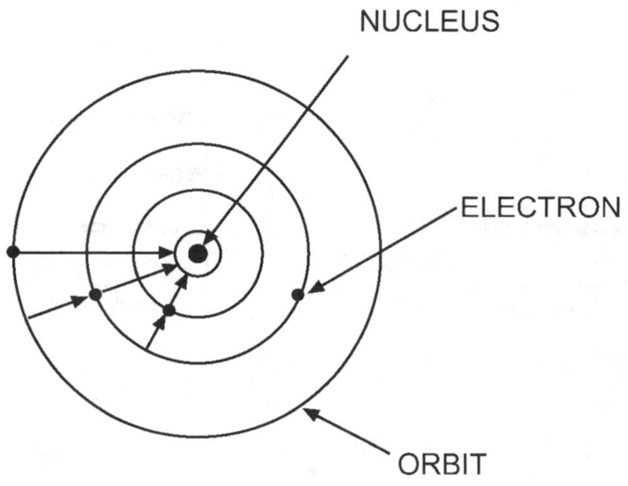

Fig. 20. Bohr's model of the atom showing the nucleus and electrons in orbit around it

---

This brings us to the name—LASER—that scientists have attached to the device. Laser is an acronym for Light Amplification by Stimulated Emission of Radiation. And we can now see why the name is appropriate. Photons of light are being stimulated into production, and amplification is occurring.

All we need for a laser, then, is a lot of atoms in excited states. At least that's the way it is on paper, but as you likely know, most things aren't as easy in practice as they appear to be on paper. A small amount of amplification may occur in a natural system of this type but not enough to make it worthwhile. The "laser effect" doesn't come that easily. We need a system in which there are lots of excited atoms, and this doesn't occur very often in nature. Most systems of atoms are in or near their "ground state." This is the lowest possible energy state of the system. If you apply heat, or other forms of energy, some of the electrons will absorb photons and jump to outer orbits. The atom is then in an "excited state."

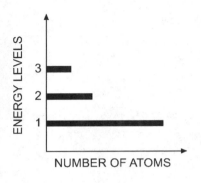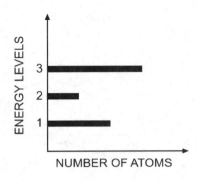

Fig. 21. The number of atoms in various energy levels showing a population inversion on the right

We can represent states of the atom by a line drawing as shown in figure 21. What we need for a laser is an excited state that has many more atoms in it than the state below it. This is referred to as a population inversion, and for many years it was an idea that few physicists dared to consider. It didn't exist in nature (under normal circumstances) and it appeared that it would be extremely difficult to create artificially. But finally in the 1950s it was accomplished. We'll look at the details of this later, but first let's go back to James Bond and *Goldfinger*.

Back to *Goldfinger*

Lasers were new when *Goldfinger* was being shot in the early sixties. But there was a lot of interest, and people were curious. Were lasers the "ray gun" we had seen in science fiction for years? Scientists were cautious, but optimistic. Many of them were convinced that lasers would eventually be capable of slicing through metal slabs several inches thick. But at the time they were barely able to penetrate a razor blade. In fact, many scientists jokingly referred to the "cutting or penetrating power" of a laser as so many "Gillettes." If the beam could penetrate one Gillette razor blade it had the cutting power of one Gillette, and at the time, lasers were only capable of penetrating a few razor blades, so they couldn't have been used to cut the gold table in *Goldfinger*. How did the producers get around that? It wasn't hard. Someone was beneath the table, cutting it with a welder's torch. If you look closely you can actually see the flames from the torch.

Incidentally, this scene is famous for something else besides the laser. The two most memorable lines in the movie were spoken in it. Everyone remembers the line, "Do you expect me to talk, Goldfinger?" And the reply, "No, Mr. Bond, I expect you to die." Goldfinger then walks away leaving Bond to the laser. Of course, Bond prevails, and he does this by asking Goldfinger, "What about operation Grand Slam?" Goldfinger, sure that no one knows about his plan, is shaken up a bit; and just before the laser reaches Bond, he has it shut off.

The cutting torch wasn't the only thing about the laser that was faked. When I use a laser in class demonstrations, one of the first things I point out is that you can't see the beam. You can easily see the red dot when you project it on a wall, but you can't see the beam itself. In the movie (and most other movies where lasers are used) you can see the beam. Aside from the fact that it's wrong, you might ask how they managed to do it. Actually it wasn't hard. They merely dubbed it in after the scene was shot.

There is a way you can see a laser beam. If you pass it through a cloud of smoke or dust it becomes visible. When I'm using a laser for demonstrations I always have a couple of chalk-filled board erasers nearby. When I clap them together, creating a cloud of chalk, the red laser beam magically appears, usually to the delight of the students. There was, of course, no cloud above Bond in *Goldfinger*.

Lasers in Other Bond Movies

As you no doubt remember, the laser in *Goldfinger* was also used to get through the main door to the Fort Knox vault. And it was also used in several other Bond movies. Only a few years later it was used in *Diamonds Are Forever*. In this movie Bond's old archenemy, Ernst Blofeld, who was posing as the billionaire aircraft manufacturer, and recluse, Willard Whyte, built a pre–Star Wars space weapon—a satellite that contained a deadly laser beam. Diamonds were incorporated in it to produce a "super laser." In the movie this laser blew up a U.S. missile and sunk a submarine—quite an accomplishment. I'll have more to say about this when I talk about weapons in space in a later section.

A laser is also used in *The Man with the Golden Gun*. The villain, Enrico Scaramanga, uses a solar-energy device to create a laser cannon, which he uses on Bond's seaplane. Needless to say, it obliterates the seaplane. In *Moonraker*, American space troops use a laser against Hugo Drax's space station. A laser is also built into the nose cone of the American shuttle that Bond uses to destroy nerve-gas spheres that are headed for the Earth's atmosphere.

I particularly liked the lasers that Bond had in the hubcaps of his Aston Martin V8 in *The Living Daylights*. He used them to slice along the bottom of a Czech police car. The beam sliced the car's body completely off the chassis (if you can believe that). With all the gadgets Bond had in his cars, it was only a matter of time before one came equipped with a laser.

Also novel was the tiny laser Bond had embedded in his wristwatch in *Never Say Never Again*. He used it to cut through some handcuffs. And finally, a laser beam was used in *Die Another Day*. The satellite *Icarus* emitted a high-powered laser beam toward the Earth to clean the minefields between North and South Korea. So, there's no scarcity of lasers in the Bond films.

More About Lasers

Earlier I mentioned that laser light is like ordinary white light, but different. Let's look at this. One of the main things that distinguishes laser light from ordinary light is that laser light is *coherent*. We know that each "bundle" of light traces out a wave, but these waves are not neces-

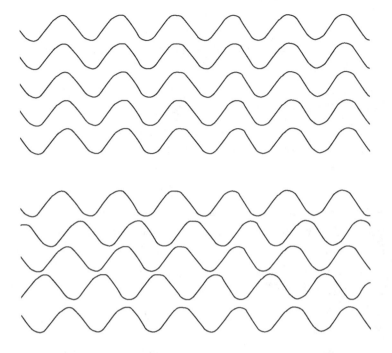

Fig. 22. Waves showing coherence. Upper ones are coherent, lower ones are not.

sarily lined up in white light. In laser light, on the other hand, they are (fig. 22). To be coherent, laser light also has to be monochromatic; in other words, it all has to be of the same frequency.

When light is coherent there is no "internal scattering" of the wavelets, or photons, as there is in an ordinary beam of white light. It can therefore be made very intense and powerful, and this has an important consequence. The best way to understand it is to think of a flashlight. Shine the flashlight into the dark. How far does it shine? Maybe a hundred feet if it is powerful enough, but, regardless of how powerful you make it, the beam won't carry very far. Try the same thing with a laser beam and you'll be in for a surprise; laser beams don't disperse (spread out), or at least their dispersion is extremely small. A laser beam was shone all the way to the moon a few years after the laser was invented, and if you think about it, this was quite an accomplishment. The moon is 240,000 miles away, and the beam spread out by only a couple of miles by the time it got there. You may also remember (if you're old enough) that when the first astronauts landed on the moon in 1969 they placed a small mirror on the surface. A laser in California reflected its beam from this mirror and determined the distance to the moon within a few centimeters.

Did this require a powerful laser? You likely think that it did, but, in reality, it didn't. We usually measure the power of a laser in *watts*. Lasers range in power from a fraction of a watt up to billions of watts. The laser used to shoot a beam to the moon had a power of only a few watts. Most handheld lasers in use today, in fact, have a power of only a few milliwatts. But some lasers have the power of hundreds of thousands of watts, and pulsed lasers give off pulses of trillions of watts over a few billionths of a second.

In summary, then, a laser beam is intense, and it can be powerful. Furthermore, lasers can vary considerably in size. With modern-day electronics we can actually build a laser that is no bigger than a grain of salt. Some of the larger ones that the military is testing as weapons, on the other hand, are the size of large buildings.

The Key: A Population Inversion

As we saw earlier, both stimulated emission and a population inversion are critical to the creation of a laser. If we have enough atoms in an excited state we can easily get a buildup of our beam. Each photon produces two identical photons (one of them is, of course, the original photon). It's like the chain reaction that occurs in the atomic bomb, but for things to go smoothly (or work at all) we need a population inversion.

How do we get a population inversion? By what is called "pumping." In effect, we pump the electrons to a higher energy level, so that the atom is in an excited state. This pumping can be done in several ways. One of the most common is to use an electrical discharge. If we are dealing with a two-level system, electrons will be pumped up to the higher level until there are more electrons in it than in the lower level.

A three-level system, where pumping is applied to the highest level, is particularly desirable in many cases. The electrons in the third level drop almost immediately to level two, and, in the process, give off photons. Charles Townes of Columbia University was the first to build a device that created a population inversion, but he did it in the microwave region of the electromagnetic spectrum, and his device was therefore called a maser (short for microwave amplification by stimulated emission of radiation) rather than a laser.

Lasers came a few years later. Townes was involved again, this time with Art Schawlow. They published a paper outlining how a laser could be built, but they weren't the first to build one. To their surprise someone beat them to the punch. Theodore Maiman of Hughes Research Laboratories in California used ruby to build a laser. Maiman's model consisted of a cylindrical ruby rod a few centimeters long and about a half a centimeter in diameter (fig. 23). One end of the ruby rod was completely reflecting and the other end partially reflecting. On being stimulated, photons traveled along the axis of the ruby, bouncing back and forth as they were reflected from the mirrors at the ends of the rod. As they traveled through the ruby they stimulated more photons until the beam was powerful enough to break through the partially reflecting surface at one end of the rod. This is the beam we see. A very intense flash lamp was wound around the rod to pump electrons to the excited state.

Other Types of Lasers

Maiman's laser was the first, and ruby lasers are still used today. But they are far from the only type of laser. Many different types now exist. One of the more common crystal lasers used today is called the YAG (short for yttrium, aluminum, garnet). YAG lasers can be made

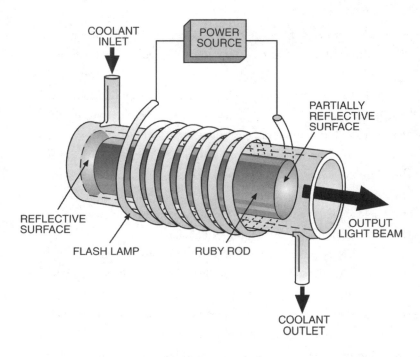

COOLANT
INLET

POWER
SOURCE

PARTIALLY
REFLECTIVE
SURFACE

REFLECTIVE
SURFACE

OUTPUT
LIGHT BEAM

FLASH LAMP

RUBY ROD

COOLANT
OUTLET

Fig. 23. Maiman's ruby laser

much more powerful than ruby lasers, and they can, indeed, drill holes through several inches of metal.

Incidentally, lasers are of two types: continuous beam and pulsed beam. In general, the pulsed beam lasers are much more powerful, but continuous beam lasers are useful in many applications, in particular, those related to medicine.

While Maiman was developing his ruby laser, Townes and Schawlow were working on a gas laser. In other words, the active material was a gas. Schawlow soon completed one in his laboratory, and several others followed. Today, gas lasers are one of the most common types. The laser I used in demonstrations was a helium-neon laser, with about 90% neon and 10% helium. A radiofrequency discharge is passed through the gas mixture to pump electrons from the ground state of helium to an excited state.

Other types of gas lasers use the rare gases argon and krypton. Argon lasers are particularly powerful and are used extensively in industry. Lasers using a mixture of the two gases have also been built. Some of the most powerful lasers, however, are carbon dioxide, and although there is no mention of it, I suspect this is the type Goldfinger was using (at least in theory). Carbon dioxide lasers can produce continuous power of hundreds of thousands of watts. They are used extensively by industry and by the military.

Another type of laser, the semiconductor laser, is now taking over much of the market. Semiconductors are materials that are intermediate between conductors and insulators. They can be either n-type or p-type; n-type materials have an excess of electrons and p-type materials, a lack of them or, equivalently, an excess of "holes." Central to a semiconductor laser is a p-n junction. With the proper arrangement and current, large numbers of electrons and holes can be generated near the junction. When an electron recombines with a hole, a photon is emitted, and with a sufficient amount of recombination, considerable light—coherent laser light—is generated at the junction. As in the ruby laser, mirrors are mounted at the ends of the semiconductor to enhance the beam.

Semiconductor lasers have many advantages. They are low power, cheap, and easy to produce. Also, they can be made very small—the size of a grain of salt, if need be. The laser in Bond's wristwatch in *Never Say Never Again* was no doubt of this type.

Liquids can also be used to generate laser action. Of this group, organic dyes are particularly important. They can be made very powerful but are expensive. Last, we have two types of lasers that are of considerable interest to the military, namely, the free-electron laser and the X-ray laser. The free-electron laser requires an electron accelerator, and a large magnet to change the direction of the beam, so it's obviously not going to be used in handheld lasers. The idea is to transform the electron beam over to a laser beam, and this is not easy, but the military has done considerable work in this area. The military are also very interested in X-ray lasers because of their potential for knocking down missiles and satellites. The problems in producing such lasers, however, are formidable, and only experimental models have been built so far.

What They Are Used For

We've seen some of the applications of lasers in the Bond movies. They were used for cutting metal and as weapons, but lasers have many other applications. The range of their application, in fact, is amazing. They are now used extensively in medicine. For example, surgeons use them to repair detached retinas, destroy tumors, repair broken blood vessels, unclog arteries, remove skin lesions, zap bleeding ulcers, or remove birthmarks and tattoos.

Lasers have also become invaluable in the communications field. Laser beams are well suited for propagation over long distances. Light has a much higher frequency than radio waves; therefore, more messages can be superimposed on laser waves. Fiber optics is now usually used in conjunction with lasers. Transparent glasslike materials can be drawn into "wires" similar to copper wires. Laser light containing the signal is fed into one end of the optic cable.

Lasers can also be seen at all checkout counters, where they are used to read the codes on the things you buy. They are also used to read CDs, in planetariums for laser light shows, and to detect pollution in our atmosphere. But we're interested in whether they can be used as they were in the Bond movies. In *Moonraker* we saw handheld laser guns, and in *Diamonds Are Forever* we saw a powerful laser in a satellite that vaporized a submarine and a missile. Let's begin with the handheld laser guns. As it happens there are a lot of problems with handheld laser guns. First, a laser of that size would not be very efficient. Much of the energy that is being generated would end up heating the gun. Second, at the present time, the laser beam from a small laser (such as those in handheld laser guns) couldn't compete with bullets. Much of the damage inflicted by a bullet is a result of its momentum, and a laser beam doesn't carry much momentum. Sure, it could burn a small hole, but to burn a hole right through your body would require about 50,000 joules of energy. Finally, building a handheld gun with that much energy is no mean task. It's possible with large lasers, but they would be too heavy to lift.

Laser guns would, indeed, have some advantages over bullets, if we could build them. Laser beams travel at the speed of light—much faster than a bullet. Furthermore, laser guns would require little "fuel" and you could use one for a long time before "reloading."

What about lasers in satellites, such as the ones in *Diamonds Are Forever* and *Die Another Day*? A laser beam travels long distances through space without spreading out very much, as we saw earlier, so it's ideal as a weapon. One of the objectives of Star Wars (or the Strategic Defense Initiative [SDI]) is to have lasers mounted in satellites that could bring down incoming missiles. X-ray lasers would be particularly valuable in this respect. Much of the work that is going on in relation to Star Wars is secret, but it's well known that the objectives of the program have not yet been met. The satellite lasers in the Bond movies are therefore still a few years away.

The Ghosts: Holograms

Closely related to lasers are holograms. If you've never seen a hologram, you should visit one (they are frequently on display in science museums). They are, indeed, amazing things. What you see looks like the real thing, but if you try to touch it, your fingers go right through it. Most amazing of all is the three-dimensional character of the image.

A hologram was used in *The World Is Not Enough* when MI6's Dr. Molly Warmflash used a hologram to show Bond how a bullet moving through Renard's (the villain) brain, was killing off his senses one by one, yet was taking a long time to kill him. The hologram of his skull was really quite remarkable. I'm surprised that this was the first time a hologram was used in the Bond movies.

What is a hologram? It looks real, but it's just a three-dimensional image. The best way to understand it is to go back to the properties of light. As we saw earlier, light has wavelength, frequency, amplitude, and phase. When images enter our eyes, we experience each of them as our brain interprets the image. Incidentally, it's important to remember that the light that is reflected from the object creates the image.

A hologram captures the three-dimensional form of the object. But isn't this what a photograph does, you ask? The answer is no. First, a photograph is only two dimensional; in short, it is sensitive only to the intensity of the light as it varies across the image. It cannot detect phase and is therefore not three-dimensional; in particular, it does not show things as your eyes see them. Your eyes are several inches apart, so you see an object from two viewpoints. In short, you experience parallax. You can demonstrate this to yourself by putting a finger up in front of your eyes, then blinking them back and forth. Notice how your finger appears to move relative to the background. This is parallax.

In holography we capture both the intensity of the wave front and its phase. This was done for the first time by Dennis Gabor in 1948, but he had problems and never actually produced a hologram. The reason is that he did not have a coherent light source, in other words, a laser. As we will see, a laser is needed for most holograms.

Two steps are needed to produce a hologram. The basic setup is shown below. In the first step a laser beam is split into two beams using a beam splitter. These beams are referred to as the object beam and the reference beam. The object beam illuminates the object, and the wave front reflected from it is recorded on the photographic plate. The reference beam comes from a different direction and is also directed at the photographic plate. Both beams must travel the same distance. When the two beams come together at the plate they don't form an image of the object, as you might expect; they form what is called an interference pattern (fig. 24).

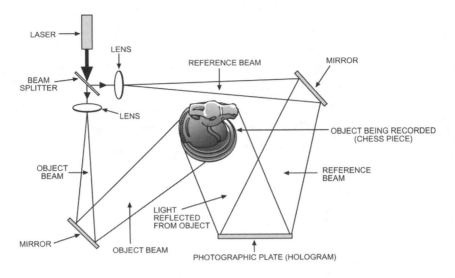

Fig. 24. Schematic of setup for the production of a hologram

Let's take a few moments to reflect on how this interference pattern is created. Consider the superposition of two waves that are exactly in phase. They will enhance one another, and the intensity of light will increase. If we have two waves that are exactly 180 degrees out of phase they will cancel one another, and there will be no light. At intermediate positions there will be partial enhancement. If a laser beam is split and brought back together, as in the case above, the overall result will be an interference pattern, which will look like a series of light and dark lines.

The hologram photograph that is created in this setup will look nothing like the object, but what is important: it contains all the information needed to create a three-dimensional image. You create this image by illuminating the photographic plate (after it is developed) with a laser beam that is directed at it from the same direction that the reference beam struck the plate. The image will be created on the far side of the photographic plate.

There are actually two types of holograms: transmission and reflective holograms. The type above is the transmission type. If the reference beam and the object beam approach the holographic plate from opposite sides, a reflection hologram is obtained. In this case white light can be used to illuminate the plate, and the image will appear in front of it.

So, that's about it for lasers and holograms. I hope this has helped you understand the lasers you have seen in the Bond movies or at least has given you an appreciation of their power, usefulness, and limitations.

How'd They Do That?

Amazing Devices

Many amazing vehicles and devices are used in the Bond movies. One of the first "amazing vehicles" was used in *You Only Live Twice*. To most people it looked like a miniature helicopter, but it was actually an autogyro, which is slightly different from a helicopter. It was called "Little Nellie," and as with Bond's other vehicles, it was well equipped with weapons. Just as amazing was the minijet *Acrostar* used in *Octopussy*; it was the world's smallest jet at the time.

Another novelty, at least at the time, was the hovercraft, a craft that moves about on a layer of compressed air. In *Thunderball* the villain Largo used a hovercraft called *Disco Volante*, and one was also used in *Die Another Day*. They are much more common now than they were in the 1960s when *Thunderball* was filmed, but they're still fascinating.

A lot of amazing devices were also used in the Bond movies, as any die-hard fan knows (and most of the die-hards could no doubt describe each device in detail). One of them was a solar device called the Solex Agitator. Used in the *The Man with the Golden Gun*, it allowed the conversion of sunlight into electrical energy with an efficiency of

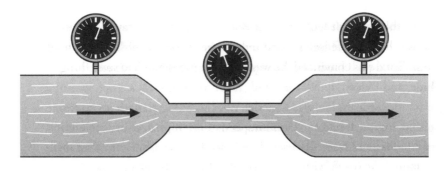

Fig. 25. Water running through a constriction. Gauges show pressure changes.

in an autogyro. When you first glance at a helicopter blade, it appears to be flat, but on closer examination you see that it isn't. It is actually shaped like the wing of an airplane, and this causes the helicopter to lift for the same reason an airplane wing causes an airplane to lift: the pressure beneath the blade is greater than the pressure above it. This can be explained by a theorem formulated in 1738 by Daniel Bernoulli.

Bernoulli showed that as the velocity of a fluid increases, its pressure decreases. Mathematically, we can state Bernoulli's theorem as

$$p + \rho v^2 / 2 = \text{constant}$$

where p is the air pressure, ρ is the density, and v is the velocity. From this equation we see that the sum of pressure and velocity is a constant (we'll neglect density for now). This means that if the velocity increases, the pressure must decrease. An interesting device called a Ventura tube shows the effect directly (fig. 25). When water rushes through the constriction it speeds up, and we see on the pressure gauges that the pressure decreases at this point. In the same way when air flows over an airplane wing, or helicopter blade, it is forced to travel faster across the top, and therefore the pressure is less there (fig. 26). This means that when a helicopter blade begins to whirl, it creates an upward lift, and

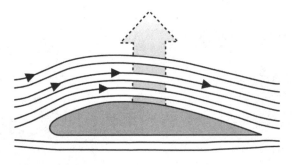

PRESSURE LESS THAN
ATMOSPHERIC

ATMOSPHERIC PRESSURE

Fig. 26. Lift on a wing due to pressure difference above
and below

---

when the lifting force is greater than the weight of the helicopter, it lifts
off the ground.

As it turns out, though, there are several instabilities in a helicopter that must be dealt with. First, we know that the outer part of the
rotor blade has a much greater velocity than the inner part. This can be
seen from the formula

$$v = \omega r,$$

where v is velocity, ω is angular velocity, and r is the distance from the
axis. Without worrying about units (they are a little difficult), let's assume the angular velocity is 10. Using the formula, we see that the velocity is $v = 10r$, and, substituting in a few numbers, we find for $r = 2$,
that $v = 20$, for $r = 3$, v is 30, and so on. It's pretty obvious from this that
the tip of the blade is flying around faster than the region near the axis.
This means that the outer region of the blade is creating a much greater
lifting force than the inner part. This must be compensated for, and it is

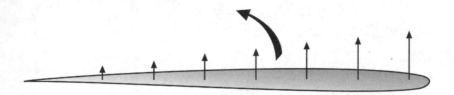

Fig. 27. Speed of a helicopter blade. Note that it moves faster with increasing distance outward.

done by giving the blade a slight twist. With this twist there is less lift on the outer section of the blade because there is less area lifting (fig. 27).

Unfortunately, it's easy to see that once the helicopter is in flight we have another problem. The blade is moving around at a constant rate, but relative to the air it is actually moving much faster when it is moving in the direction the helicopter is going (as compared with the opposite direction). To illustrate this, let's assume the helicopter has a velocity of 100 mph. The blade is, of course, rotating, but at any position it also has a linear velocity that can easily be calculated. Let's assume it is 50 mph. When the blade is moving forward (in the direction the helicopter is moving) we must add it to the helicopter's velocity. This means the blade is actually moving 150 mph relative to the air. When the blade is moving in an opposite direction its velocity relative to the air will be 100 – 50 or 50 mph. This means that the lifting force is much greater in the direction the craft is flying, and this causes the fuselage to tilt. How do we adjust for this? If you look closely at the point near the axis where the rotor blades are attached you will see a mechanism that allows them to "flap." In short, they tilt when moving in the forward direction so that less air strikes them (as compared with when they are moving in the opposite direction) (fig. 28). This is referred to as "blade flapping."

Another problem is related to what is called torque. Torque is defined mathematically as

$$\tau = Fd$$

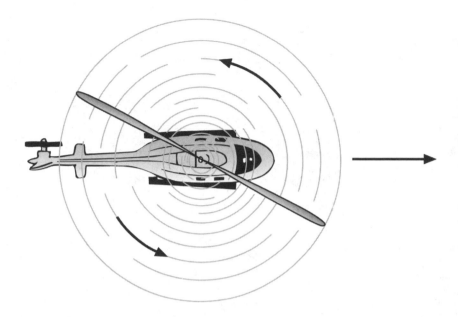

Fig. 28. Note that the speed of the forward-moving prop is greater

where τ is the torque, F is the force, and d is the perpendicular distance from the axis, or fixed point, to the line of the force, as shown in the diagram (fig. 29). Because the spinning blade causes a force, it also creates a torque, and according to Newton's third law, for every action there is an equal and opposite reaction. This means that the whirling blade causes the fuselage of the helicopter to rotate in a direction opposite to the direction in which it is rotating. This can be compensated for in two different ways. First, a small propeller that creates a torque that balances the torque caused by the rotor is placed on the tail. Second, two rotor blades are used on the helicopter, with one rotating clockwise and the other rotating counterclockwise.

Finally, there's one last problem. Before I discuss it I have to talk about how a helicopter moves once it gets in the air. After all, the large rotor blade only creates an upward lift. Basically, you have to tilt

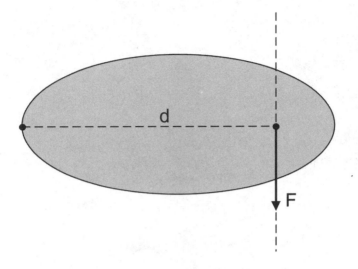

Fig. 29. An illustration of torque. Torque is force times the perpendicular distance from it to the axis.

the helicopter so that part of the thrust from the rotor blades is in the forward direction. But here we encounter what is called the gyroscopic effect. It is well known that when a force is applied to a rotating body, the reactive force is not in the same direction; it is at ninety degrees to the direction of the applied force. This must also be taken into account.

The World's Smallest Jet

In *Octopussy* a piece of top-secret U.S. military equipment fell into Cuban hands and Bond was assigned to destroy it. It was held at a Cuban air base. Bond entered Cuba disguised as a Cuban officer, but his disguise didn't help for long. He was soon captured, but as he was being taken to a Cuban jail, sexy CIA agent Bianca pulled alongside the army truck and distracted the driver. She was pulling a horse trailer, and in the trailer, camouflaged as a horse was *Acrostar*—a minijet (fig. 30). Bond got loose, jumped into the minijet, and escaped. The Cubans

Fig. 30. *Acrostar*

quickly fired a heat-seeking missile at him, and for a while it appeared that he wasn't going to get very far. In a last desperate move, however, he headed for the hangar where the secret piece of military equipment was stored. He flew *Acrostar* through the large hangar door. The guards tried to close the back door before he could exit, but *Acrostar* was too quick. It got through, but the missile chasing it blew up the hangar and, in the process, destroyed the piece of equipment Bond was assigned to destroy. Mission accomplished!

    Acrostar was the world's smallest jet at the time. It was powered by a TRS-18 microturbo jet engine, had a maximum dive speed of 350 mph, and was amazingly maneuverable. It needed a takeoff run of 1,800 feet, but could land in 800 feet. It was 12 feet long and had a wingspan of 13 feet. In the movie, Bond ran out of fuel but managed to land; he pulled up to a pump at a gas station where he ordered a station attendant to "fill it up."

    Jet engines are amazing, and they work because of Newton's third law. One of the easiest ways to see this law in effect is to blow up a small rubber balloon and let it go. As I'm sure you know, it will fly off in a dizzying array of loops and flips. As the air makes its way out of the balloon, it forces the deflating balloon in the opposite direction.

    The first jet engine was designed in 1930 by the English RAF

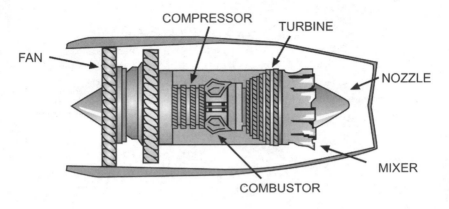

Fig. 31. Details of a jet engine

lieutenant Frank Whittle. It was not tested until 1937, however, and did not fly until 1941. About the same time a German physicist, Hans von Ohain, also began working on a similar design, but it wasn't until after the war that production really took off. General Electric and Pratt and Whitney both soon had jet engines that worked.

Several different types of jet engines are now common. We'll begin with what is called the turbojet (fig. 31). In principle, the turbojet design is relatively simple. At the front is an inlet that allows air to enter; from here it is compressed by using blades that squeeze it into decreasing volumes. The air that exits this region has 30 to 40 times the pressure it had when it entered, and the increase in pressure increases the air's temperature to over a thousand degrees. Some of this air then enters what is called a combustor. Jet fuel is squirted in the combustor, where it vaporizes and mixes with the air. It is then ignited, which causes it to heat even more dramatically. It leaves the combustor at a temperature of about 3,000°F.

Some of the air leaving the combustor turns a turbine, which consists of several blades. It is the turbine that supplies the power to the compressor. The remaining air is blasted from the nozzle, creating a reactive force that pushes the turbine (and airplane) forward. In many cases there is also a mixer behind the turbine that combines the high-temperature air with cooler air that didn't go through the combustor. In addition, an afterburner is usually near the rear of the turbojet that helps create extra thrust. Extra fuel is sprayed into the gases in the afterburner. And, finally, there is a fan in the front.

A propeller can be attached to the main shaft; this gives what is called a turboprop. The turbine in the rear, in effect, turns the propeller, which in turn, propels (or partially propels) the aircraft. Turboprops are used in some smaller airliners and in transport crafts; they work best at speeds less than 500 mph.

The ramjet is a simplified version of the turbojet that is used in guided missiles and space vehicles. It has no moving parts and can't work alone; a high speed—usually greater than the speed of sound—is required for it to operate. At this speed the air is rammed into the chamber with such a force that it doesn't need to be compressed or heated. The ramjet therefore has no turbine or combustor.

The Hovercraft

Another amazing vehicle, the hovercraft, was used in two Bond movies. It was first used in *Thunderball*. Largo's yacht, the *Disco Volante*, contained a hovercraft. The front section was a hovercraft with a twelve-cylinder engine (it had a horsepower of 1,200); the back section was an attached cocoon that was equipped with several large guns. After the underwater fight between the Navy Seabees and Largo's men, Largo jettisoned the cocoon and sped off in an attempt to escape. But Bond managed to cling to one of the hydrofoils and eventually make his way into the main cabin where he fought with and eventually overcame Largo.

A hovercraft was also used in *Die Another Day*. Near the beginning of the movie Bond was in a hovercraft being chased through a minefield between North and South Korea. It was an exciting and action-filled chase, but in the end Bond was captured and spent the next few months in a North Korean prison.

A hovercraft has three main components: a basic platform, a motorized fan, and a skirt (fig. 32). The fan supplies the air that is blown between the platform and the ground or water. The skirt helps keep it contained. Furthermore, the air flowing into the chamber forms a ring of circulating air around the main mass of air, which also keeps it from escaping. The chamber is referred to as the plenum chamber (from the Latin word "full"). The air in the plenum chamber is at a much higher pressure than the surrounding air and it cushions the platform, allowing it to "float." For this to work, the pressure pushing upward has to balance the force of gravity acting downward.

Pressure is force per unit area, or

$$P = F/A,$$

where P is pressure, F is force, and A is area. For example, assume we have a hovercraft that weighs 3,000 pounds (including the people on it) and is 20 feet long by 10 feet wide. It has an area of 200 square feet, and if the hovercraft is to hover, the air cushion below it must exert an upward force of 3,000 pounds. This means the pressure of the air must be

$$P = 3,000/200 = 15 \text{ lbs/ft}^2,$$

which is not a particularly great pressure.

We have to steer the hovercraft, of course, and it obviously can't be steered like a car, since it has no contact with the ground. In most hovercrafts a fan is attached near the rear, but in some cases the fan that supplies the air for the air cushion is used. Some of this air can easily be directed out the back of the craft, and it will exert a reactive force according to Newton's third law. Finally, a rudder is also used to give the craft stability.

Hovercrafts can be used equally well over water or ground. It is essential, however, that the ground be relatively level and even. Hovercrafts are incapable of climbing even small hills.

DEATH RAYS, JET PACKS, STUNTS & SUPERCARS

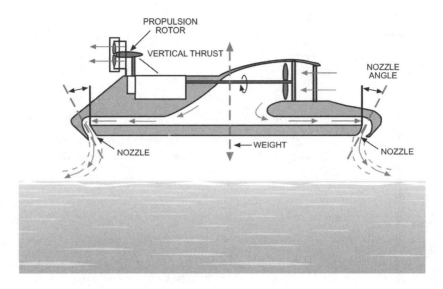

Fig. 32. Details of a hovercraft

An obvious advantage of the hovercraft is the absence of friction between it and the ground, because it sits on an air cushion.

Energy from the Sun . . . And It's Free!

Solar energy was featured in *The Man with the Golden Gun*. The villain, Francisco Scaramanga, possessed a solar device called the Solex Agitator that was capable of converting solar energy into electrical energy with 95% efficiency. This is far beyond what we are capable of today. Scaramanga used the Solex Agitator to power his hideaway on an island near Phuket in Thailand. His solar energy complex was capable of generating a lot of power, but it was never clear what he intended to do with all that energy. He would obviously have a lot left over after he powered everything on the island. He did, however, have an interesting "solar gun" that he used to destroy Bond's seaplane. Energy from such a system is usually stored in large batteries, but Scaramanga had low-temperature vats for storing his energy, and, strangely, he had only one technician to oversee it.

A solar device was also used in *Die Another Day* in the orbiting satellite *Icarus*. It contained a solar mirror that emitted a laser-like blast of sunlight across the Korean demilitarized zone, burning up thousands of landmines and, in the process, incinerating trees and killing wildlife. It had a diamond-based reflector technology that was capable of turning night into day, or of channeling the sun's rays into a narrow, energetic, destructive beam.

The sun produces a lot of energy. By Earth standards, it is indeed incredible, and we use only a small fraction of it. Ninety-nine percent of the sunlight that reaches the ground is converted to heat and radiated back into space. If only a tiny fraction of this energy could be captured and converted to usable energy (such as electrical), all the energy demands of the Earth could easily be met. The sun is an "inexhaustible" supply of energy—unlike fossil fuels, it will never be depleted (at least not for millions of years). The problem is converting it to electrical energy efficiently.

The major device for converting sunlight to electrical energy is the photovoltaic cell. The photovoltaic cell came about as a result of discoveries made in 1887 by Heinrich Hertz and later by the German physicist Philipp Lenard, who was Hertz's assistant. In 1902 Lenard began studying what is now called the photoelectric effect. He showed that an "electrical effect" was produced when light was shone on certain metals; in other words, electrons were released at the surface of the metal by the light. Einstein explained this strange phenomenon in 1905. According to Einstein, light was made up of particles (called photons) that released electrons when they struck the surface of the metal. In particular, the number of electrons released depended only on the frequency of the light, and not on its intensity or brightness, as might be expected. This was an important breakthrough, which later won Einstein the Nobel Prize.

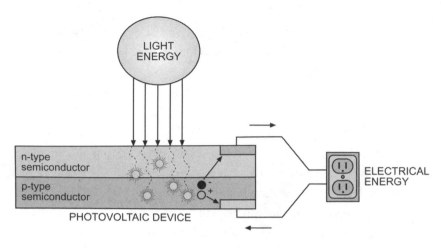

Fig. 33. A photovoltaic cell

The photocell does not generate electricity by itself. A voltage must be applied to it because there is nothing to attract the electrons when they are freed from the metal. In a photovoltaic cell, semiconductors supply this voltage (semiconductors are materials with an electrical conduction intermediate between that of insulators and metals). They conduct electrons reasonably well, but not as well as certain metals such as copper and silver; furthermore, they can be made into two varieties, referred to as n-type and p-type. The n-type semiconductors have an excess of negative charge, the p-type semiconductors have an excess of positive charge.

When a layer of n-type semiconductor is placed next to a layer of p-type semiconductors, we get what is called a p–n junction (fig. 33). In a junction of this type we get an electric field. When photons of light strike the surface of this p–n junction it releases electrons. If an external circuit is attached to the junction as shown in the diagram, the electric field will cause the electrons to flow through the wire to the other side of the junction, generating an electrical current. The amount of electricity that can be generated in a single photovoltaic cell (or solar cell) is, as you might expect, relatively small. So in practice many solar cells have to be linked together.

Scaramanga's solar cells were assumed to be about 95% efficient. In reality, in laboratory settings, we can achieve efficiencies of only about 25 to 30%. Commercial cells, in general, have efficiencies of only about 16%. A lot of work has gone into increasing this efficiency in recent years, and we anticipate that much more efficient cells will become available. At the present time, however, it is still difficult for solar energy to compete with fossil-fuel energy. There are a lot of places, though, where it is much more feasible to use solar energy. One is in satellites; a solar cell was first used in 1958 in the Vanguard satellite, and it has been used in satellites ever since. Much of the TV you watch is beamed around the world by satellites powered by solar cells, and small calculators and many toys use solar cells.

As we saw earlier, single photovoltaic (PV) cells generate only a small amount of power, typically only a few watts (the watt is a power unit obtained by multiplying current by voltage). In practice, PV cells are usually connected into solar panels. They can generate up to 60 watts and more and create voltages of 110 volts, which is what we usually use. Such solar panels are commonly used in isolated areas, but more and more people are incorporating them into their houses, and they're also used extensively in RVs.

Watch Out! You're on Radar

Everyone worries about getting caught in a "radar trap" and having to pay a speeding ticket. I'll admit that I've been caught more than once. Radar is common today. Not only is it used by police to catch speeders, but it is also used by the airline industry to keep track of their flights, by weather forecasters, and by the military. Incidentally, radar is an acronym for radio detection and ranging.

Radar is featured in several of the Bond movies. I don't think

Bond ever got caught in a radar trap for speeding, even though in most movies he went well over the speed limit. But how exciting would it be if he always obeyed the law? Anyway, we're not going to say much about speed traps. Our interest in radar is in how to avoid it, or more exactly, how to be invisible to it. The Stealth Ship in *Tomorrow Never Dies* was designed to avoid detection by radar, and the spaceship in *Moonraker* was supposedly impervious to radar.

What exactly is radar? From a simple point of view, when the police catch you on radar, they use a "radar gun" that emits a beam of electromagnetic waves that strikes the front of your car. Some of these waves are reflected back to the gun, which analyzes them and determines how fast you are going. Determining how far you are away at any given time is relatively easy since electromagnetic waves travel at the speed of light (186,000 miles per second [mps]). This means that they take $1/186{,}000 = 0.0000053$ second or 5.3 microseconds (μsec) to go a mile. To get the distance from the radar gun to the target we need to double this value because the signal goes to the target and back. This gives

$$\text{Distance} = (\text{time to target})/10.6 \ \mu\text{sec}.$$

As an example, assume a radar gun receives an echo 5 μsec after it was transmitted. The distance of the target is $5/10.6 = 0.47$ miles. If we determine this distance a few seconds later we can determine the target's speed, since we know how far it traveled in a given time.

Actually, a better way to do this exists, and it's based on what is called the Doppler Effect. The Doppler Effect is a change in frequency (or wavelength) that occurs when a body emitting (or reflecting) waves is either approaching or receding from an observer. It applies to all types of electromagnetic waves and to sound waves. In the case of electromagnetic waves, if the object that emits or reflects them is approaching us the frequency of the returning waves is decreased. We say that it is shifted toward the shorter-wavelength end of the spectrum; we refer to it as blue-shifted. If it is moving away from us, it is shifted toward the longer-wavelength end, or it is red-shifted. We can easily determine the speed of the object by noting how much it has shifted. A device for calculating and analyzing such shifts is built into police radar guns.

But we're interested in how objects avoid radar. There are many ways of doing this. Let's begin by looking at what types of electromagnetic waves make up radar. Electromagnetic waves range all the way from radio waves, through microwaves, infrared, visible, ultraviolet, X rays, and gamma rays, but most of these rays are not appropriate for radar. Radar beams are usually in the microwave range.

Now, let's consider the methods that are used to avoid radar. First, not all radar waves are reflected from an object. The percentage that is reflected depends on the absorption properties of the surface of the object, and any object can be covered with radar-absorbing material. Dark surfaces usually reflect less than bright surfaces. This is why airplanes that want to avoid radar are painted black. Second, all surfaces don't reflect in the same way. If we want to avoid radar we want a surface that reflects as few rays as possible back to the detector. In other words, we want a surface that scatters the rays extensively; in this case, few will get back. Surfaces that are particularly good at this are triangular-shaped surfaces (or W shaped). This means we will want to make our object with a triangular surface, as much as possible. Finally, we can send out a signal to interfere with the returning signal. This is also helpful.

Airplanes and other vehicles that want to avoid radar have what is called a radar cross section (RCS). A bird, for example, has a RCS of approximately 0.01 square meters. The B2 bomber, on the other hand, has a RCS of 0.75 square meters, which is quite small and would therefore be hard to detect by conventional radar. The F-22 Raptor has a RCS of approximately 0.01 square meters.

I'm not sure what the RCS of Elliot Carver's Stealth Ship was, but it was presumably no larger than a submarine periscope, so it was small. Also the ship had a black hull, which was made of graphite, and it was shaped to avoid radar.

X-ray Glasses: Are They Possible?

Bond used X-ray glasses at Zukovsky's casino in *The World Is Not Enough* to check who was carrying a gun. The glasses also saw through clothes. Indeed, you occasionally see ads in magazines for "X-ray glasses" that see through clothes. Is this really possible?

Like ordinary light, X rays are part of the electromagnetic spectrum. They are much shorter in wavelength (and higher in frequency) than ordinary light waves and are therefore much more energetic. Our eyes are sensitive only to visible light, so we can't see X rays directly. X rays were discovered by the German scientist Wilhelm Roentgen in 1895. He found them by accident while experimenting with vacuum tubes, and within a week he had used them to take a photograph of his wife's hand. To his surprise, it showed the bones in her hand and her ring clearly, but the flesh on her hand was not visible. Roentgen called the new rays X rays, because he had no idea what they were.

As I said, the human eye is not sensitive to X rays. So how do we see them? We don't see them directly, but we do see their results. If we could see X rays we would only see things that emitted them or stopped them; indeed, this is what an X-ray film sees, and we can observe this film. The film shows the "shadow" of things that X rays can't travel through. They don't travel through bone and metal, so we see it in the body.

Now for X-ray glasses. Somehow, X rays would have to be generated by them. In an X-ray machine they are generated by suddenly stopping high-speed electrons, so we would need something in the glasses to accelerate the electrons and stop them; glasses aren't big and complex enough for this. Besides, even if the glasses could generate X rays, we wouldn't be able to see them directly. I hope this convinces you that there is no such thing as X-ray glasses. So you can save your money the next time you see an ad. But I've seen pictures taken through glasses of this type! you say. Indeed, some types of goggles allow you to see into the infrared and some types of materials do allow infrared rays through them. Your body, however, does not, so it might appear that you are seeing through clothing.

A Bullet-Deflecting Magnetic Field

Another interesting device Bond used was a watch that generated a very intense magnetic field. It was used in the movie *Live and Let Die.* Referred to as a "hyper intensified magnetic field," it was powerful enough to deflect a bullet at "long range." Bond didn't use it to deflect any bullets, although I can't see why . . . he was always getting shot at. He did, however, use it to whisk a spoon off M's coffee saucer, and he later used it to unzip a woman's dress. What is strange is that if it was powerful enough to deflect bullets, why didn't it attract every metal object near it when it was turned on? It appeared to be strangely selective. And if it could deflect bullets, why didn't Bond use it in other movies?

Again we ask, is it actually possible to have a magnetic field, generated by a wristwatch that is strong enough to deflect a bullet? Let's begin by looking at magnetic fields. A magnetic field is, of course, generated by a magnet. This magnet can be a permanent magnet, or what is called an electromagnet, which uses a coil of wire to generate the field. A magnetic field is, in many ways, like the gravitational field, except that it attracts certain metals instead of massive objects. The magnetic field falls off in strength in the same way; in other words, it drops off inversely as the square of the distance. This means that a magnetic field falls to one-quarter of its value when the distance is doubled.

We usually measure the strength of a magnetic field in gauss. A small, hand-held magnet might have a field of a hundred gauss; powerful magnets used in the lab can have fields of 100,000 gauss or more.

As you likely know, magnets have both a north and a south pole. If you cut a magnet in half, each half will again have a south and a north pole. In fact, you can continue this indefinitely, so you obviously can't isolate a given type of pole. We also know that two like poles will

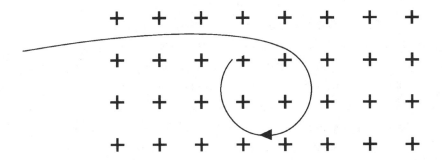

Fig. 34. A charged particle in a magnetic field. Note that it traces out a circle.

repel one another, and two unlike poles will attract one another. But what about a particle, or small object, in a magnetic field? Will it be attracted or repelled? If the particle is charged, it will indeed be affected by the magnetic field. Let's assume it comes in perpendicular to the field lines, as in figure 34. It will experience a force given by

$$F = qvB$$

where q is the charge of the particle, v is its velocity, and B is the strength of the magnetic field.

This force will push the particle perpendicular to the field so that it ends up tracing out a circle. It does this for both positive and negative charges; the only difference is that they will go around the field lines in opposite directions.

But a bullet is not charged. Furthermore, it's much more massive than a tiny charged particle. What will happen in this case? We know that magnets do affect certain metals. They attract them. Does this mean a magnetic field would attract a bullet? We know that magnets attract iron; indeed, they actually attract several metals including iron, cobalt, nickel, and gadolinium; these materials are referred to as ferromagnetic. If you could closely examine their structure you would see that they are made up of millions of tiny magnets, which we refer to as magnetic domains. These tiny magnets are usually randomly oriented; in other words, their north and south poles point in arbitrary directions. But when they are brought into an external magnetic field these tiny magnets all line up, creating their own field. In fact, they will always line up so that they are attracted to the source of the external field.

Most magnets used in devices are electromagnets. Electromagnets work because their magnetic domains line up in response to an external magnetic field. A current generates a magnetic field around it, so if you wrap wire around a ferromagnetic material such as iron and pass a current through it, you generate a magnetic field (it will generate a magnetic field even if there is no ferromagnetic material). This magnetic field causes the domains in the iron to line up, thereby enhancing the overall field. The magnetic field generated by Bond's watch would have to be of this type.

With this information we can now analyze the magnetic field generated by Bond's watch. First, a wristwatch is too small to create an extremely strong magnetic field. Most strong electromagnets now use high-temperature superconductors or liquid helium cooling, which would be difficult to incorporate in a wristwatch. Second, the bullet from a gun is not charged, so it would not be deflected by a magnetic field. Even if it were, however, it is so massive, the effect would be negligible. And third, bullets are usually made of lead, which is not affected by a magnetic field. True, some bullets have steel shields, but they would have far too much momentum to be deflected, even by the strongest fields. Also, "at long distance" is a problem, because magnetic fields fall off rapidly with distance.

Electromagnetic Pulses and E-Bombs

The devastating effect of electromagnetic pulses is featured in at least two of the Bond films. It is a central feature of *GoldenEye* and is also mentioned in *A View to a Kill*. The GoldenEye is a disk that programs two satellites to fire an electromagnetic pulse that destroys all electrical equipment, but presumably does not kill people. Strangely, an electromagnetic pulse was directed at the Russian outpost of Severnaya in *GoldenEye* and it appeared to devastate everything and killed everybody except for two people.

Electromagnetic pulses (EMPs) and the devices that generate them are now a large concern of the military. Considerable work has been done in the United States and Russia and also in smaller countries such as India. The history of the EMP actually goes back to 1945. Even before the first atomic blast was set off, the physicist Enrico Fermi tried to calculate the magnitude and effects of the electromagnetic pulse that would be generated by the atomic blast. He was only partially successful, and it wasn't until the early 1960s that scientists realized how devastating this pulse could be. During a high-altitude nuclear test over Johnson Island in the Pacific, scientists determined that the electromagnetic pulse generated by the blast had severely affected the Hawaiian Islands, which are 850 miles away. Streetlights and fuses failed on Oahu and telephone service was completely disrupted for several hours on Kauai. Furthermore, the blast seriously affected the instrumentation of an airplane 200 miles away. The Russians also noticed similar effects.

At the time, little was known about EMPs, but in recent years the physics has been studied in detail, and we now know quite a bit. And what we have learned is scary, to say the least. For years our main worry in an atomic blast was the blast damage, the danger of the thermal radiation, and the radiation effects. These are, indeed, destructive, but the first two are significant for only a few miles around the blast center. Radiation, however, can be picked up by the winds and circulated around the world, and its effects can be significant (e.g., cancer). The EMP does not affect humans directly, so few worried about it at first.

What causes an EMP? In a nuclear blast, large fluxes of gamma rays (very high energy x rays) are generated. They, in turn, produce high-speed electrons, some of which can become trapped in the Earth's magnetic field. The surging electrons produce extremely high electric and magnetic fields that, when coupled with electrical and electronic systems, produce huge currents and voltages that destroy the equipment. In particular, the pulse is so powerful that any long metal object would act as an antenna and generate high voltages as the pulse passed. And there is a large array of such objects on Earth: electric wires, fences, railroad lines, metal beams in buildings, et al.

All electrical equipment is destroyed by an EMP, including all computers, communications, telephones, car electrical systems, and so on. Such a blast would bring a society to a standstill and cause billions of dollars in damage, even if it didn't endanger human life. What is really worrisome is the huge impact a single pulse would have. Anything in the line of sight is affected. This means that a large device detonated about 250 miles up over a central state in the United States, such as Kansas, could knock out the electrical systems and electronic devices in all 48 contiguous states. This would be a disaster. Every electrical device in the nation would be blown up, and every car on every freeway in the United States would come to a stop. It's hard to believe, but it's true. Furthermore, it would be a major job getting each of them running again.

An EMP would make a great offensive weapon. We could use it to disable enemy troops without killing them. Because an EMP affects the electronics in ballistic missiles, we could set one off and destroy the electronics in incoming ballistic missiles. Of course, we would have to be careful not to harm our own weapons, and that would be difficult.

Does any defense against such a pulse exist? Yes. Electronic systems can be shielded, but it is costly. Many military devices are now shielded, but it would be too costly to shield all civilian electronics.

So far we have talked only about an electromagnetic pulse that is generated by a nuclear blast. You don't need a bomb, though, to generate such a pulse. It is relatively easy to build what is called an E-bomb. One of the simplest types is the flux compression generator (FCG); it consists of a tube of explosives that is surrounded by coils of copper wire.

Just before the explosives are detonated, current is sent through the coil, which creates a magnetic field. The explosion causes a short circuit that compresses the magnetic field and creates an electromagnetic pulse that is just as powerful as one created in a nuclear blast.

In many ways, although it does not kill people, the E-bomb would be much more devastating to society than a nuclear bomb.

Bond's Gadgets and Gizmos

M r. Bond must be cold after his swim. Put him somewhere warm," said Drax.

Jaws pushed Bond as he led him to the chamber directly beneath the rocket jets. Goodhead was already in the room. She rushed up to him when the door opened.

As they held one another the roof slid back. Standing near the edge was Drax. He raised his hand mockingly, "I bid you farewell." The roof closed and he headed for the Moonraker rocket.

Bond looked quickly around the room. Spotting an air vent, he ran over to it. Taking out his watch, he attached it to the bars, then pulled a wire out of it and told Goodhead to get behind him. An explosion rocked the room, blasting the vent open. Bond and Goodhead jumped through the opening and sprinted up the narrow passageway as the rocket blasted off.

Again, one of Bond's gadgets saves him. In almost every movie he uses some sort of gadget. Many of them are quite ingenious, and everybody is fascinated by them. The first gadget appeared in *From Russia with Love*. Q gave Bond a briefcase that contained a knife, 50 gold sovereigns, 20 rounds of ammunition, a .25 caliber AR-7 folding rifle, and an exploding canister of gas. Needless to say, they were helpful.

The train scene was one of the most exciting in the movie. In it Bond was knocked out by the villain, Red Grant. When he regained consciousness, Grant was sitting over him with a gun in his hand.

"Red wine with fish," said Bond. "I should have known."

Bond tried to sit up. "What lunatic asylum did they get you out of?" he asked.

Grant struck him across the face with his gun. " . . . crawl across the floor . . . and kiss my feet," he said as he screwed a silencer on his gun.

Bond eyed him cautiously. "Can I have a last cigarette?" he asked.

"No," replied Grant.

"I'll pay for it," said Bond. "There are gold coins in the briefcase."

Grant got the briefcase down and tossed it to Bond. "Open it."

Bond opened it and handed him the strips of coins.

Grant looked up and saw another briefcase. He asked if there were any coins in it.

"I would imagine so," said Bond. "Let me take a look."

Grant grabbed the case. "I'll open it," he said.

Slowly he undid the latch and as it opened the case exploded in his face. Bond lunged forward and the two men began fighting. But Grant had a secret weapon: a garrote in his watch, and he quickly used it on Bond.

Just as it seemed that Bond was going to black out, he reached down and grabbed the knife from the briefcase. He stabbed Grant, disabling him, then grabbed his garrote and strangled him.

Gadgets helped Bond out again, but, surprisingly, in this scene Grant also had a gadget, the garrote that was hidden in his watch.

Q: The Gadget Master

The Secret Service's equipment officer, Major Boothroyd, or Q as he is better known, was responsible for Bond's gadgets; and he always did an excellent job. We're never quite sure what to expect, but in every movie there's always an amusing scene where Q presents a new gadget to Bond. In each scene Bond begins playing with something in the lab as Q starts to explain how the device works. Q gives him a dirty look and says, "Now pay attention 007." Occasionally he even yells, "Be careful!" as Bond toys with one of the gadgets. Despite his apparent inattention, though, when it comes time to use the gadget, Bond knows exactly what to do.

Many of the scenes were quite humorous. In *A View to a Kill*, for example, Q demonstrated his pet dog "Snooper." It was a robot with a retractable neck and eyes that contained a video camera, allowing Q to monitor the area around the dog. He also demonstrated a fake arm cast that sprung forward with devastating force, a chair with an ejection seat, a gun that fired cement, and some exploding bolas. Q eventually became quite popular, and Desmond Llewelyn, who played Q, was surprised by all the fuss fans made over him.

Bond always made good use of the gadgets that Q gave him, and in most cases they saved his life. In addition, they were something that fans looked forward to and enjoyed. As far as I'm concerned, they added a lot to the movies.

Tick, Tick, Tick . . . Geiger Counters

Geiger counters are used in two Bond movies: *Thunderball* and *The World Is Not Enough*. They are devices that detect radiation and are used extensively by technicians and others that work around radiation.

In *Thunderball*, two atomic bombs are hijacked and Bond is sent to locate them. He ends up in Nassau in the Bahamas with Emilio Largo as a prime suspect. On inspecting the bottom of Largo's yacht, Bond finds a secret door and suspects that the bombs may be in the hull of the yacht. He gives the heroine, Domino Derval, a camera with a Geiger counter in it so she can check it. The counter begins to click, but before she can do anything, Largo discovers what she is up to and drags her to her cabin and ties her to the bed.

In *The World Is Not Enough* nuclear physicist Christmas Jones has a radiation detector attached to her belt, and she is seen carrying a large case that contains instruments and various types of nuclear equipment. This case likely contains a Geiger counter.

Hans Geiger invented the Geiger counter in 1913 shortly after radioactivity was discovered. His first model was crude but improved significantly in 1928 when he teamed up with Walther Müller. The device could detect three types of radiation: alpha, beta, and gamma rays. Alpha particles are nuclei of helium atoms, which are two protons and two neutrons that are bound together. They are weak and can easily be stopped by a few sheets of paper; they are not dangerous. Beta rays are high-speed electrons. They are much more penetrating than alpha rays and can pierce about a quarter inch of aluminum. Gamma rays are the only true rays. They are similar to X rays but have an even greater penetrating power; it takes about two inches of lead to stop them. All three rays are given off by radioactive materials; all three are emitted in an atomic blast.

The Geiger counter is actually a relatively simple device. It consists of a gas-filled metal cylinder that acts as one electrode of a circuit (fig. 35). A thin wire is placed along the center of the cylinder; it acts as the other electrode. A voltage is applied that is just below the threshold needed to cause a current to pass through the gas from one electrode to the other. When the device is brought near a radioactive substance, alpha, beta, and gamma rays knock electrons off the atoms of the gas molecules and they become charged. As a result they are attracted to one of the electrodes and cause a small, brief current to flow. This current is amplified by an electronic device so that an audible click is heard. The gas quickly returns to its normal state, allowing each click to be registered. The number of clicks per second gives a measure of the intensity of the radiation.

Geiger counters played a large role in the study of radioactive substances, but they were largely superceded in 1947 by what is called a halogen counter. Halogen counters work on the same principle but

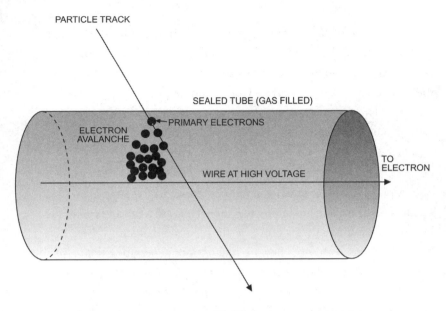

SEALED TUBE (GAS FILLED)

PRIMARY ELECTRONS

ELECTRON
AVALANCHE

WIRE AT HIGH VOLTAGE

TO
ELECTRON

Fig. 35. A Geiger counter showing an incoming particle

have a much longer life and a lower operating voltage. More recently many different radiation detectors have been invented, including scintillation counters and various solid-state devices.

Voice Boxes and Sound

In *Diamonds Are Forever* Bond was amazed that Blofeld was able to change his voice by using a tiny box. He had taken over billionaire recluse Willard Whyte's empire and was running it from the top of one of the casinos; he was giving commands to Whyte's subordinates by using a device that made his voice sound like that of Whyte. Bond was so impressed that, with the help of Q, he used a similar device to trick Blofeld. He wanted to know where Blofeld was keeping Willard Whyte so he phoned him, pretending to be his henchman, Bert Saxby. Blofeld fell for it and told Bond where Whyte was being held.

How does a voice box such as this work? Because we're dealing with sound we must begin with its basics. According to physics, sound is a pressure wave or disturbance that travels through a medium at a certain speed. The speed of sound in air is 343 m/sec. The source of the sound is a vibrating material of some type—the vocal cords of a person in the case of a voice, or the vibration of the strings in the case of a musical instrument.

These vibrations travel through the medium as a wave. As we learned earlier, there are two types of waves: transverse and longitudinal. Sound is a longitudinal wave; in other words, it vibrates in the direction that it propagates. A given sound is characterized by its frequency, or pitch, and its amplitude. Frequency is a measure of the number of vibrations per second that are occurring and it is measured in vibrations per second or Hertz (Hz). The human ear can hear a range from about 20 Hz up to about 15,000 Hz. In the musical scale, middle C has a frequency of 264 Hz and G has a frequency of 396 Hz.

One of the many problems of growing older is that the upper end of this range decreases significantly as you age. Older people therefore do not hear frequencies anywhere near 15,000 Hz. I used to conduct an interesting experiment in class that showed this quite clearly. Using a sound generator that was capable of producing a full range of frequencies, I would tell the students to put their hand in the air if they could hear the sound. Then I would gradually increase the frequency and tell them to lower their hand when they could no longer hear it. It was always interesting to see the results, and it was always the older students who put their hand down first. Needless to say, there were always many hands in the air long after I could hear anything. Incidentally, dogs can hear much higher frequencies than humans can.

The amplitude of a wave is associated with its intensity, which in the case of sound is its loudness. The human ear is also capable of hearing a large range of intensities. We usually measure intensity in terms of decibels (dB). Rustling leaves have a decibel reading of about

10 dB. Normal conversation is usually about 60 dB (assuming there is no arguing). The threshold of pain is 130 dB, and, interestingly, this is not much higher than the front row at a rock concert, which is about 110 dB. At 160 dB your eardrums will be perforated.

I have an ongoing argument with my son about the dangers of loud sounds and the damage that it can do to the eardrum. Whenever I get into his truck the radio is usually on rock music at about 100 dB. I love music . . . but, there is a limit. One of the consequences of too much loud sound is tinnitus—a ringing in the ear. I learned about tinnitus firsthand a few years ago after using power tools too often without ear protection—and it was no fun. Fortunately, it went away.

Getting back to the voice box—how would you duplicate somebody's voice? As you know, everyone's voice sounds a little different. You can easily identify individuals on the phone by the sound of their voice. The two major differences in voices are frequency and overtones. We can easily distinguish a woman's voice (with a few exceptions) by its frequency; it is usually higher pitched. More characteristic of a person, however, are overtones; they are present in almost every sound. Assume, for example, that you sound middle C on the piano and analyze it with an oscilloscope (an instrument that puts waveforms on a screen). You might expect it would be the same as the sound generated by a frequency generator that is dialed to 256 Hz (such instruments give a "pure" sound). But it isn't. Comparing the two waves, you see distinct differences. The wave for the piano is different because of overtones. Indeed, if you sounded the same note on a violin it would also look different (fig. 36).

Overtones, frequency differences, and the way different people pronounce different words distinguish one voice from another. A voice is just as characteristic of a person as his or her fingerprints. In the voice boxes used in the Bond films, the characteristics of a person's voice would have to be carefully analyzed, then superimposed on the voice of the speaker (or word substitutions could be made). To see how this might be done let's look at how a voice is recorded, transmitted, and picked up. First, a microphone is used to pick up the sound of the

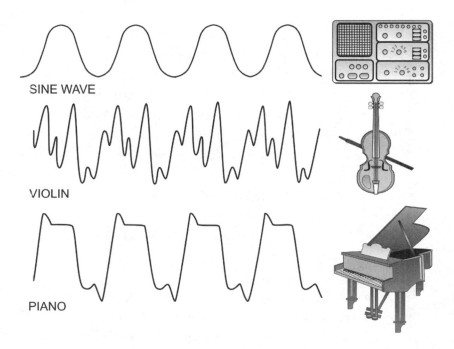

SINE WAVE

VIOLIN

PIANO

Fig. 36. Middle C sounded on a signal generator, a violin, and a piano

voice; in this microphone, pressure variations of the sound wave force a diaphragm to vibrate. These vibrations are converted to an electrical current with a resistor or small voltage generator, and the resulting current is applied to another diaphragm or speaker at the other end of the circuit. It is relatively easy to modify an electrical signal, and this is what is needed to change the voice. In the voice boxes used in the Bond film, the characteristics of the speaker's voice would have to be carefully analyzed and the electrical signal changed accordingly. In practice, a computer would be needed to do this. The first step would be to look at the signal on the computer's screen; here it would look like a series of jiggles or waves. These "jiggles" can easily be modified by using the controls on the computer. To change a voice so that it sounds like another voice you would have to begin with the voice you wanted to duplicate. You could build up a vocabulary of words, then substitute the appropriate word when it is encountered, or you could actually change the form of the wave corresponding to each word so that it conforms to the other person's voice. This would probably be more difficult. Regardless of which method you used it would be a tedious and difficult process; it would have to be done automatically and quickly. Musicians do this sort of thing routinely. They analyze the waveforms of their recordings on a computer screen and change them according to what they want. They can easily replace incorrect notes and mispronounced words, and they eliminate extraneous noise, but they don't try to make their voices sound like someone else's (at least I hope they don't).

Today, we can record voices digitally; therefore, changing them is much easier than it was a few years ago. In fact, many different types of voice changers are now on the market, ranging from inexpensive ones that only change the frequency to expensive professional models that also change the harmonics. People sometimes use them on their telephone when they don't want to be identified. Even with the more expensive models, however, it would still be difficult to duplicate anyone's voice accurately . . . which is perhaps a good thing.

Polarized Light

In *A View to a Kill* Bond used special glasses that eliminated glare and gave him the ability to see through the reflections on a window. In particular, he saw Zorin paying Stacy for a transaction of some sort and he became suspicious.

Sunglasses that reduce glare are now quite common; they are known as polarized sunglasses, and a large percentage of the sunglasses now sold are polarized. When I'm talking about polarization in my classes I occasionally ask the class if they prefer regular or polarized sunglasses. As you can likely guess, most people say they prefer polarized glasses. But when I ask them why (after all, they cost more), they're not sure. "You can see better with them," is the usual reply.

Indeed, under certain circumstances, you can see better.

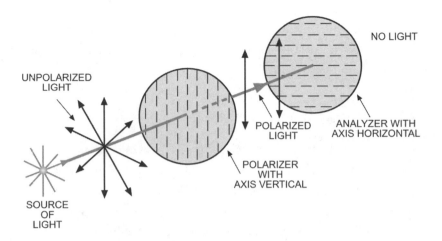

Fig. 37. Light passing through two Polaroids with axes perpendicular

Polarized sunglasses do eliminate glare, and it's relatively easy to show how. If you could examine a light beam closely you would see that it is made up of many "wavelets," all vibrating perpendicular to the direction of propagation, but at random angles. Polaroids are translucent materials that allow only the light rays through that are vibrating in a particular direction, namely the direction of the axis of the Polaroid. Assume we have two of these Polaroids, one lined up behind the other as in figure 37. Assume that the axis of the first Polaroid is vertical; this means that only waves vibrating in the vertical direction will get through. Now, turn the axis of the second Polaroid so that it only allows waves through that are perpendicular to the axis of the first one. But there are no waves vibrating in this direction; they were eliminated by the first Polaroid, so no light gets through. When I demonstrate this in class, students are surprised that the Polaroids are black when these axes are perpendicular, then as I turn them so the axes line up or are parallel, they become translucent.

How does this affect sunglasses? Well, whenever light is reflected from a surface—a window or a pool of water—it becomes polarized in a direction parallel to the surface. If the Polaroids in your sunglasses have their axis perpendicular to this, you will see no glare. Fishermen particularly like polarized sunglasses because it allows them to see into the water and see the fish.

Magic Wristwatches

Bond wore a lot of fascinating wristwatches in his movies. And I'll have to admit, I'd love to have some of them. One of the first of these watches was seen in *Live and Let Die*. We have already looked at its most interesting feature—it deflects bullets. It also had a buzz saw in it that allowed Bond to cut ropes that were binding his hands.

In *The Spy Who Loved Me* Bond had a Seiko that printed out a ticker tape. It was used in the opening sequence, at a very inappropriate time for Bond. The message was from M and it told Bond to report immediately. He also had a Seiko in *Moonraker*, and it allowed him to escape a fiery death beneath the Moonraker rocket as it took off.

Bond's watch in *For Your Eyes Only* doubled as a walkie-talkie. It was used at the end of the film when Bond was being congratulated by Prime Minister Thatcher for doing such a good job. The Prime Minister didn't know, however, that she was talking to a parrot. Bond was otherwise occupied. Bond also used a Seiko in *Octopussy*. A homing device had been put in a Fabergé egg, and the watch contained a receiver that allowed Bond to track the egg.

The most impressive Bond watches came in the last few movies. In *GoldenEye* Bond used an Omega that was equipped with a laser, a Geiger counter, a miniature saw that could cut through glass or steel, a telex communicator, and a cellular phone. A watch with this many gadgets that actually worked would have to have been much larger. Nevertheless, it was impressive. He used a similar Omega in *The World Is Not Enough*. It had a laser beam, a powerful flashlight, and a grappling gun. So much for watches.

Miniature Cameras

Miniature cameras are used in two Bond movies. In *Moonraker* Bond used a small camera to photograph the Moonraker plans that he took from Drax's safe. A humorous touch was added; when he was photographing the plans you could see a 7 after the two circular lenses — giving 007. A ring camera was used in *A View to a Kill*. Bond used it to take photographs of some of the people at Zorin's estate, one of whom was a former Nazi physician who performed experiments on concentration camp inmates.

How does a camera work? It's actually a relatively simple device — at least in principle. The lens of the camera forms an image of the object on a film at the back of the camera. Surprisingly, a lens isn't even needed; a pinhole would work, but the image it produces is not very sharp, and its light-transmitting power is low.

With a single lens, deviations from the "ideal" image usually occur. In good cameras these errors are corrected for by using other lenses. Some of these errors are chromatic aberration (a color dispersion of the light similar to what occurs in prisms), spherical aberration (the deviation of rays so that some are brought to focus at different points than others), astigmatism (horizontal and vertical lines in an object are brought to focus in different planes), and coma (a comet-like elongation of the points outside the center of the image). All these things can be corrected for by various lenses, and this is why all good cameras have complex lenses consisting of several individual lenses.

The image also has to focus on the back of the camera, which requires additional lenses. For a given object distance we therefore have to know where the image forms. This is given by the formula

$$1/O + 1/I = 1/f$$

where O is the object distance, I is the image distance, and f is the focal length of the lens.

For the photographer, two settings are of particular importance. Indeed, without the correct settings you won't get a good picture. These settings are the aperture size and the shutter speed. The aperture is the opening into the camera body; it controls the amount of light that reaches the film. Too much light and the film will be overexposed; too little, and it will be underexposed. Aperture sizes are measured by f-stops. Standard sizes are f/1.4, f/2, f/5.6, f/8, f/11, f/16, f/22, f/32, f/45, and f/64, where f/1.4 has the largest opening and f/64 has the smallest. Shutter speeds are measured in seconds and range from 1, 2, 4, 8, 15, 30, 60, 125, 250 through 500, where 1 is one second and 60 is 1/60th of a second.

Most of the cameras that Bond used were miniatures; in fact, one was a ring camera. Miniature cameras have been around almost since cameras were invented, and they have always been somewhat of a novelty. A watch camera was made as early as 1886. The technology took a leap forward in the 1930s when Walter Zapp developed the Minox camera. His film was only 9.5 mm across. The 1950s and 1960s, though, produced really small cameras. None were as small as a ring, however. According to Guinness World Records, the smallest commercial film camera ever built was 1.14 inches wide by 0.65 inches thick, which is considerably larger than a ring. You undoubtedly know, however, that a rather dramatic development has occurred in relation to cameras in the past few years. Film has been superceded by digital cards; in other words, images are now stored digitally, rather than on film. This has allowed cameras to become much smaller, so that ring cameras are now common. You now see them advertised everywhere.

Global Positioning System

GPS is all the rage today, with millions of units now in use around the world. GPS stands for Global Positioning System, and it's a technique that allows you to locate yourself very accurately on the surface of the Earth (fig. 38). It all started in 1978 when the Department

Fig. 38. A global-positioning system (GPS) unit

of Defense launched the first NAVSTAR satellite. The object of the launch was to allow the military to navigate land, sea, and sky to a very high precision. It wasn't long before civilians got into the act. They began using the system in 1980.

There are now 24 GPS satellites in the sky (if you include those of Russia and Europe, there are many more); this is the number needed for GPS to work. By the early 1990s the system had really taken off, and it is now used to guide ships, boats, and land vehicles of various types.

A GPS unit was used in *Tomorrow Never Dies,* but it was a little different from the units I have been discussing. It was called a GPS "encoder" and could manipulate the signal from the NAVSTAR satellite that was used by others to get their location. The terrorist, Henry Gupta, bought it at the "weapons market" at the beginning of the movie. He used it to change the signal from NAVSTAR so that it threw the position of HMS *Devonshire* off by about 70 miles. As a result the *Devonshire* was in Chinese waters and Chinese MIGs were sent out to warn it. One of the MIGs was shot down and the *Devonshire* was sunk by Carver's Stealth Ship in an attempt to force England and China into a war. Bond came to the rescue, and war was averted.

Small, handheld GPS units are now also used extensively by hunters, hikers, and other outdoor enthusiasts. They give your position on a map with amazing accuracy, and they have no doubt stopped a lot of people from getting lost. Speaking of getting lost . . . yes, I have to admit I was lost once and would have liked to have had a GPS unit, but unfortunately they weren't around at the time. I was hunting in a rather dense, forested area along a twisting river. After getting separated from my hunting companion I realized that I was seeing the same landmarks for a second time and began to get worried.

I knew that if I walked perpendicular to the river for long enough I would eventually get out of the trees and be able to see where I was. So I started off, making sure I was going in a straight line. Eventually I saw a clear area ahead, and thinking I was finally out of the woods I began to run. To my surprise I saw that the clear area was the river. I couldn't believe it—I was sure the river was behind me. Oh well, I thought . . . I'd try it again, and I did. And to my surprise the same thing happened. This was crazy. Was I really going around in circles? I finally gave up that strategy and just walked. Soon it began to get dark and I resigned myself to spending the night out there. I knew it would get cold.

Suddenly, to my relief, I saw somebody coming down the trail toward me. Rather sheepishly I told him I was lost and asked him if he knew the area very well. He chuckled and said that he had hunted it for

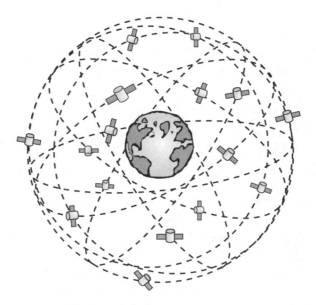

Fig. 39. Satellites needed to locate a GPS receiver on Earth

years. He asked me where my car was parked. I described the area to him as well as I could. "I think I know where it is," he said finally. He gave me instructions on how to get there, and I followed them as best as I could in the dark. It was pitch black by the time I got to my car. My hunting companion had called the sheriff and they were standing near the cars talking as I approached. I was quite embarrassed but relieved.

Let's turn now to how GPS works. Basically, GPS is a radio-based worldwide navigation system composed of 24 satellites (fig. 39) and associated ground stations that uses a slight variation of triangulation (called trilateration) to calculate the terrestrial location of an observer. The process starts with three satellites, each of which emits a signal that spreads out around the satellite in the form of a sphere. The receiver on the surface of the Earth will be at the intersection of the three spheres. The distance to each of the satellites can easily be determined by using the time the signal takes to get to the satellite (the signals travel at the speed of light). Timing obviously has to be extremely accurate, and there is no problem with the satellites; they each have atomic clocks. But the GPS receivers on Earth have much less accurate clocks, and this causes problems. To compensate, measurements from a fourth satellite must be taken.

The two major problems for GPS users are the Earth's ionosphere, which can delay or interrupt the signal from the satellite, and the electrical interference near the surface of the Earth. Engineers are working to overcome these problems and much more accurate GPS units are expected soon.

Q in Las Vegas

Of all the devices that Q made there's one I would like to have more than any other. In *Diamonds Are Forever* he used what he called an "electromagnetic rpm controller" that caused slot machines to pay off every time. Even Desmond Llewelyn said that, of all the devices Q made, he wished this one had actually worked. I'm not sure how it was supposed to work, but I'm surprised it didn't catch somebody's attention in the casino (besides Tiffany's). Even if you could control the rate of spin, or rpm's, I'm not sure how you could stop it whenever you wanted, but it's a great idea.

A device like this would obviously be a nightmare for the casino owners, but I don't think anyone would feel too sorry for them. And I don't think they have to worry. With my luck at the casinos, though, I could use such a device. The only time I ever won a large jackpot (and it was only nickels) was the first time I put a coin into a machine, and that was a long time ago. So many nickels came out I didn't know what to do with them all. Ah . . . if that could only happen a little more often.

Spotting the "Bugs"

What was the first thing Bond did when he got to his room each night, particularly if it was a new room? If you're an ardent fan, you know that he searched for "bugs" (and I don't mean the four-legged variety). He usually found a few. Occasionally he even had gadgets to help him in his search. In *A View to a Kill* his shaver contained a detector that he used to look for bugs at Zorin's estate. And in *Live and Let Die* he had a bug detector in his hairbrush. In *From Russia with Love* he didn't need any gadgets to find the bugs. He found so many in his room that he rang the front desk and asked for another room. They told him that no others were available—except the bridal suite. "It will do," he said, but surprisingly, he didn't check it for bugs.

Bugs were used a lot during the cold war; embassies around the world bugged each other. Today, bugging is still used extensively by industry and it has reached a high level of sophistication. Early bugs were usually tied into phone lines or placed in a room to monitor conversations. They could be hard-wired or listened to from a short distance away via a radiofrequency (RF) signal.

RF detectors are still used a lot, but many of the detectors used today are more sophisticated. An RF signal coming from a bug can be detected easily. Because of this, infrared, microwave, and laser bugs are now commonly used. They are generally more difficult to detect. A laser beam, for example, can be bounced off a window from a distance, and because the window is vibrating in accordance with the conversation going on inside, the vibrations can be amplified and converted to audible sound. In practice, this technique is not highly efficient, but it works. Snoopers using lasers and other such devices can now sit in a car half a block away and monitor what is going on in a room.

Fiber optic lines are now used extensively for transmitting signals, and there are numerous signals on any given line. Nevertheless, snoopers have also been able to break into them.

Amazing Cellular Phones

Cellular phones are everywhere today. Sometimes it seems that half the people walking down the street have a phone glued to their ear.

The sight of people roaring down the highway at 90 mph with a phone at their ear always sends a chill down my back.

None of these phones are as sophisticated as some of the phones Bond has used. One of the most amazing Bond phones was the Erickson mobile phone used in *Tomorrow Never Dies*. Besides being a phone, it was a remote control for Bond's BMW 750iL, it had a fingerprint reader, and a key duplicator (how often would he need that?), and a laser beam capable of cutting steel; it was also a 20,000-volt stun gun. Quite a phone in anybody's book.

Bond could start his car from a distance by tapping twice on a pad in the phone. He could then move it by drawing his fingers across the pad. In particular, he could bring it to where he was hiding. He could also drive it from a concealed position in the back seat by using the video screen in the phone; the screen showed the view through the front window. His phone also alerted him to hazards such as missile attacks.

Bond had a similar device for controlling his BMW Z8 in *The World Is Not Enough*. In this case it was a key ring. And like the Erickson phone it had a starting pad and an accelerating and braking pad, plus many extras.

We're one up on Bond, though; as far as I know, he never took a picture with a cell phone, and picture-taking cell phones are all around us now. You can, in fact, take a picture and send it to somebody over the phone (as if I need to tell you that).

Underwater Gadgets

The first underwater gadgets appeared in *Thunderball*. One of the most interesting was an underwater breather that had 4 to 5 minutes of air and could be used in an emergency. Bond used it twice: first in Largo's shark pool and later in the extended underwater battle. The British military was so impressed with the device that they approached the filmmakers to find out how it worked. They were disappointed to find that it was a fake. Bond was actually holding his breath when he was pretending to use it. The same breathing device was used in *Die Another Day* when Bond swam under the ice.

One of the world's first underwater cameras was also used in *Thunderball*. This might seem surprising, because underwater cameras are now common and have been around for quite a while. Bond used the camera in *Thunderball* to photograph the hidden doors in the hull of Largo's yacht. The camera used infrared film and could be used in the dark.

An underwater "jet pack" was also used in *Thunderball*. With it strapped to his back, Bond could swim at a high speed underwater. The jet pack was armed with spear guns and a searchlight. Bond used it in the underwater battle.

Safe Decoders

As a highly prolific spy, Bond obviously had to break into a few safes. He didn't take money, of course; his interest was usually in the plans for something or other. His first safe breaking comes in *On Her Majesty's Secret Service* when he uses a rather cumbersome safe-cracking device. It is so large that it has to be hoisted up to the high-rise office of Blofeld's lawyer with a crane. It presumably had a magnetic sensor in it. With it, Bond found out that Blofeld was attempting to claim rights to the title of Count, and he also found out that Blofeld's new hideout was at Piz Gloria in Switzerland.

A safe decoder was also used in *You Only Live Twice*. Bond breaks into Osato's headquarters in Tokyo and uses a state-of-the-art device to steal documents from his safe. This unit is much smaller than the one used in *On Her Majesty's Secret Service*. It worked by running through every possible safe combination at high speed until it found the right one.

A similar device was used in *GoldenEye*. In this case an elec- yboard adhered to the lock and automatically dialed all pos- nations in a matter of seconds, until it got to the right one. gain access to the Archangel Soviet nerve gas facility.

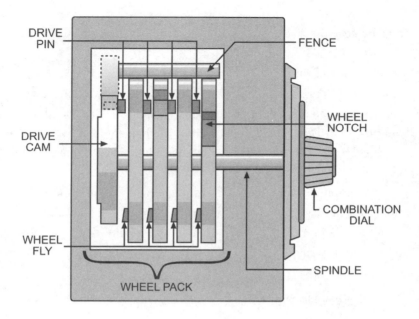

Fig. 40. Side view of the wheelpack of a safe combination showing the drive cam.

How do safes work, or more exactly, how do safe combinations work? If you take a combination apart you will see that it contains several wheels or cams that are usually made of steel. Each has a notch, or indentation, in it at the outer edge, and each has two teeth or "pins" on it, one on each side, as shown in figure 40. The set of wheels is referred to as the *wheelpack*. The number of wheels in the wheelpack is equal to the number of numbers in the combination; in other words, there is one number for each wheel.

The dial of the combination is attached to a rod that passes through the center of the wheels but is not connected directly to them; it is connected to what is called the drive cam, which sits at the re of the wheelpack. The drive cam also has a pin on it, and when it this pin eventually makes contact with the pin on the first v begins to turn. The pin on the opposite side of this whee

DEATH RAYS, JET PACKS, STUNTS & SUPERCARS

contact with the pin on the next wheel and it begins to turn, and so on, until all the wheels are turning in unison.

As we saw, each wheel has a notch on its outer edge. When the proper numbers are dialed (the combination) the notches are all lined up. Adjacent to the drive cam is a lever with a bar or "fence" attached to it. This fence normally keeps the safe locked, but when all the notches are lined up, it falls into the groove that they create and allows the door to be opened.

Although Bond didn't use the "lock-picking" technique used by many safecrackers, it's an interesting technique. As we will see, it is more complicated than usually depicted in the movies. The safecracker needs a lot of patience, time, and a good ear; many safecrackers, in fact, don't have a "good ear" so they use a stethoscope. They have three main obstacles to overcome. The first has to do with the drive cam; like the wheels, it also has a notch in it, but it is different from the wheel notches, which are usually square. It has sloped edges so that the catch on the end of the lever arm (and attached fence) can easily pass through it. The safecracker must determine exactly where this notch is, and how wide it is. He does this by listening. When the catch at the end of the lever hits one edge of the drive cam notch it makes a slight click; he has to listen for this click, and for a second click at the other edge. From them he can determine the position and size of the notch (its size is referred to as the contact area).

Next, he has to determine how many wheels there are in the wheelpack. Simple safes have three, but most have more (up to six or eight). To do this he starts with the dial pointed at a position directly opposite the contact area. He then moves the dial slowly to the right and listens carefully. Each time the drive cam picks up a new wheel there is a slight click. If he hears four clicks, for example, and no more, he knows there are four wheels, and therefore four numbers in the combination.

So far, so good, but now the real work starts, and this is the tedious part. The safecracker begins by resetting the lock by turning it to the right several times. Then, starting at zero, he turns the dial slowly to the left, listening again for clicks corresponding to the two sides of the contact area. He notes their position on a graph, then moves the wheel three numbers to the left of zero and repeats the process. The clicks will come at a slightly different position this time, and he records them on his graph. He continues this for the entire wheel, and when he's finished he has a graph where points are concentrated around certain numbers. If there are four wheels, there will be four of these numbers. They give the combination of the safe, but you still don't know what order they should be used in. You have to try all possibilities, but one of them will work.

This is the sophisticated way of getting into a safe, and it's easy to see why most safecrackers don't use it—it takes too long. There are other methods, of course, but they're much messier. You could drill through the safe into the region where the wheelpack is located, then use a viewer to line up the notches in the wheels. In most cases this requires a little luck. Or you could use a welding torch and cut your way in. But there's a problem with a large amount of heat; you could easily cook everything inside before you got to it. Besides, it takes quite a bit of skill to operate the torch properly. Finally, you could blow the door off its hinges using nitroglycerine. But the contents of the safe might not survive the explosion, and such an explosion is likely to attract a lot of attention.

Bond, of course, never used any of these methods. The last three were a little too uncouth for him, and the first may have been too tedious. He used a computer device that ran through all possible safe combinations at high speed until it found the right one. Is it likely that the small device he had could do this? A simple calculation shows that the number of possible combinations is incredibly large—far beyond the capabilities of his small device.

Homing Devices

Homing, or tracking, devices play an important role in two of the Bond films. The first homing device appeared in *Goldfinger* when

Bond used one to follow Goldfinger to his Swiss factory. Bond attached a small transmitting device to the underside of Goldfinger's Rolls Royce after he played golf with him. He kept track of it by using a direction finder on the dashboard of his Aston Martin.

A homing device was also used in *Octopussy*. It was placed in a Fabergé egg and gave off a signal that was detected by a radio receiver in Bond's pen and a direction indicator in his watch. He used them to track Kamal Khan to his palace near Delhi, India.

There is nothing new or mysterious about homing devices. They are merely small transmitters that give off a signal at a particular frequency. Whoever is tracking the device uses a receiver tuned to the frequency (a direction finder may also be used). With older devices you had to be close enough to pick up the signal, but now, with the advent of GPS, the art of tracking has changed dramatically. With GPS you can track almost anything, including your spouse, children, pets, older relatives, and your vehicles. You can, for example, strap a wristwatch-like device on your child, and if he disappeared it would be relatively easy to find him. Stolen cars are now tracked with similar devices. So they're obviously quite helpful.

The Three-Dimensional Visual Identograph

A rather fascinating device called a 3D visual identograph was used in *For Your Eyes Only*. Bond and Q used it to identify the underworld figure Emile Locque. Locque was the man who paid off the Cuban hit man Hector Gonzales for murdering Timothy and Iona Havelock, Melina's father and mother. Gonzales gunned them down on board their yacht while they were searching for the wreckage of the *St. Georges*.

The identograph contained a screen on which Bond made a rough sketch of Locque. Along with Q, he gradually refined it until it resembled Locque quite closely. This is, of course, what is done by police sketch artists. What was different was that the machine had a database of thousands of pictures and files of felons; it was able to tap into the CIA, Interpol, the

The Incredible Bond Cars

A h, the Bond cars! What would a Bond movie be without fast sports cars? Many of us can easily picture ourselves behind the wheel of a fast sports car like the ones Bond drives. I'll have to admit I'm not immune. Of course, most of us can't afford the kind of cars we see in the Bond movies. We can dream, though, and I've been doing that for a long time. My father was a garage owner, so I was brought up around cars, and I can remember the excitement I felt when a "flashy" new car was brought into the showroom.

We all like to compare the various cars. Open any car magazine and you'll see pages filled with specs on new cars. It seems as if the manufacturers are all trying to outdo one another. Some of the things you see are horsepower, torque, displacement, compression ratio, acceleration from zero to sixty, and top speed. Let's begin with them.

A Brief Survey of the Physics of Cars

Many of the basic concepts of physics are used in describing the properties of cars—in particular, how powerful they are and how fast they go. Bond's cars are always powerful and fast. A car's power depends on the size of its engine, which in turn depends on the size of the cylinders in the engine, and how far the piston moves. We refer to these as the *bore* and *stroke* of the engine. Both are given in units of either inches or millimeters, where one inch is 25.4 millimeters (mm).

Here are some examples.

	Bore (in./mm)	Stroke (in./mm)
2003 Audi A4 Cabriolet	3.25/82.5	3.65/92.8
2004 Mercedes-Benz CLK 320	3.54/89.9	3.31/84
2003 Saab 9-3 Arc	3.38/86	3.38/86
2003 Infiniti FX 45	3.66/93	3.26/82.7

As the piston moves back and forth in the cylinder it displaces a certain volume. We refer to this as the *displacement* of the engine. Actually, the displacement we usually refer to in relation to cars is the total displacement of all the cylinders. It is obviously a good measure of the size of an engine and is usually given in liters, cubic centimeters (cc), or cubic inches (ci). The conversion factor between these units is 1 liter = 1000 cc = 61 ci. The displacements of a few modern engines are as follows.

2005 Chrysler PT Cruiser	2.4 liters
2003 Infiniti FX 45	4.5 liters
2004 Volkwagen Touareg	3.2 liters
2004 Pontiac Aztec	3.4 liters
2004 Jeep Liberty	2.4 liters
2004 Hummer H2	6.0 liters

It's easy to see that most vehicles range between about 2 and 6 liters, with six being on the high side. The only thing you usually see in this range are trucks or large SUVs. Indeed, it's rare to see anything higher than about 8 liters.

Let's turn now to one of the most important terms in physics, namely, *work*. We know that cars perform work so it's a term that applies to them. Work is related to energy because it is energy that produces it, and the energy that produces the work comes from the gasoline in the car. We define work as the applied force times the distance the force acts through. It therefore has units of ft-lbs. We can rewrite force × distance as (force/area) × (volume) which is pressure × volume, or

$$W = PV,$$

where P is pressure and V is volume. This is the definition we usually use when talking about work in relation to cars. The pressure and volume we are referring to here is, of course, the pressure in the cylinder and its volume.

We have to be a little careful, though. This definition is what is referred to as *indicated work*, W_i, because it is not the work that is delivered to the crankshaft. Because of friction and so on, the actual work is less; we refer to it as *brake work*, W_b. The ratio of indicated to brake work is referred to as the *mechanical efficiency*, E_m, of the engine

$$E_m = W_b/W_i.$$

Modern cars typically have mechanical efficiencies of the order of 75% to 95%.

This brings us to two of the most common terms used in relation to cars. Everyone has heard of torque and horsepower. Let's begin with torque. As we saw earlier, torque is defined as force times a lever arm, where the lever arm is perpendicular to the direction of the force. Since we are dealing with a length and a force you might think that the units would be ft-lbs, but this is the unit we used for work, so to avoid confusion we revise the order when we talk about torque. The units of torque are therefore lbs-ft.

Since we have a force and a perpendicular lever arm, torque is obviously a measure of twist, or twisting effort. It doesn't produce any straight-line motion, but it's important in relation to cars, since an engine is undergoing rotary movement (or twist). Torque is a measure of how powerful this rotary motion is. There's a slight problem, though. The torque produced by a car depends on its rpm's; in short, it peaks at a particular rotary speed, so we have to specify the rpm's when we talk about torque.

With this in mind let's look at the torque of a few modern vehicles. All are 2004 models.

Chrysler Pacifica	250 lbs-ft @ 4,000 rpm
Cadillac SRX V8	315 lbs-ft @ 4,400 rpm
Porsche 911 GT3	284 lbs-ft @ 5,000 rpm
Dodge Viper SRT10	525 lbs-ft @ 4,200 rpm
Volkswagen Passat W8	273 lbs-ft @ 2,750 rpm
BMW X5	324 lbs-ft @ 3,600 rpm

For some reason the torque values of Bond's vehicles are rarely mentioned, but they do mention the horsepower of all of them, so let's turn to that.

We'll begin with the concept of *power*, which is the rate of doing work. Since the units of work are ft-lbs, the units of power are obviously ft-lbs/sec. In 1783 the Scottish engineer James Watt decided this was a rather obscure unit; he wanted something that people could visualize a little more easily, so he began experimenting with how much work a horse could do in a given amount of time. He determined that one could raise a 150-lb weight 4 feet in one second; using this he defined the horsepower (hp) to be 550 ft-lbs/sec.

If you look up the horsepower of any of the Bond cars, you will usually see it specified as braking horsepower (bhp). Because this is the actual power delivered to the crankshaft, it's the one we should be using. In the future, therefore, when I talk about horsepower I will actually be referring to bhp.

The horsepower of a few modern cars is as follows (all are 2004 models, unless specified otherwise). Again we have to specify that the horsepower also varies with rpm's.

Dodge RAM 2500 HD (truck)	345 hp @ 5,400 rpm
Lexus RX 330	230 hp @ 5,600 rpm
BMW 645 Ci	325 hp @ 6,100 rpm
Chrysler 300C	340 hp @ 5,000 rpm
Jeep Liberty	210 hp @ 5,200 rpm
2003 Toyota Matrix XR5	180 hp @ 7,600 rpm

We see that the range is from about 150 to more than 500. The high-powered cars we see today are amazing. Only a few years ago a car with a horsepower of 300 was rare. Today, it's relatively common. It's a good idea to keep this in mind when we look at some of the older Bond cars. They were extremely powerful for their day, but look pretty ordinary by today's standards.

One of Bond's cars was a turbo, so we'll briefly discuss turbocharging. And, of course, along with turbocharging comes supercharging. To discuss turbos we have to begin with what is called *volumetric efficiency*; it is the efficiency with which the fuel-air mixture is brought into the cylinder. Suppose the cylinder has a volume of 120 cc when the piston is at the bottom. You would think that the engine would draw in 120 cc of fuel-air and fill the cylinder to capacity. But it doesn't. The amount of fuel-air coming in may be considerably less than the volume of the cylinder (fig. 41). The volumetric efficiency is a measure of what fraction is actually filled, and as it turns out, it is usually about 80 to 95%.

If we could fill it up all the way we would obviously get more power out of the car, and engineers have managed to do this by compressing the fuel-air mixture as it enters the cylinder. We can use the exhaust to drive a turbine that compresses the mixture, or we can compress it more directly with a compressor that is driven by a pulley off the crankshaft. The former is referred to as turbocharging, the latter as supercharging.

Finally, I should mention a few other measures that are usually used in comparing cars, and they are used in comparing the Bond cars. They are the acceleration time from 0 to 60 mph, the top speed of the

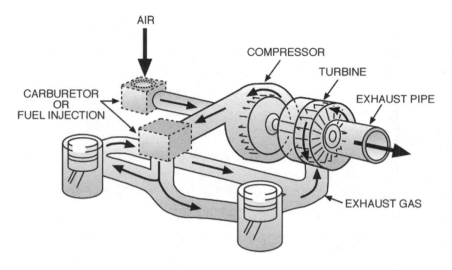

Fig. 41. Schematic of a turbocharger

car, and the time to brake from 60 to 0. Braking doesn't really apply to the Bond cars, though, because Bond never brakes—he only accelerates. We will also occasionally mention the type of brakes on the car and the type of suspension system.

Aston Martin DB5

Sports cars did not play a large role in the first few Bond movies. Bond drove a Sunbeam Alpine in *Dr. No*, which was rented from a woman in Jamaica for 15 shillings (about $3.50) a day. In *From Russia with Love* he drove a Bentley, mostly because Ian Fleming had driven a Bentley and liked them. It was not until the third film, *Goldfinger*, that Bond finally got a car worthy of him. Fleming had given Bond an Aston Martin in the book, which he equipped with a few "extras." The producers, however, felt that Bond needed much more. It's interesting, though, that when the special effects expert for the movie, John Stears, approached the Aston Martin car company, he informed them that he would like to use one of their Aston Martins in the film, but EON

Productions couldn't afford to actually buy one. Things have obviously come a long way since then. Anyway, Aston Martin arranged to loan them a car, and they selected the most powerful one they had—the DB5. Ken Adams, the production designer, and Stears then had to decide how they wanted to equip the car. And, wow! Did they ever equip it! It soon became the most famous car in the world.

The car had front-mounted machine guns, concealed behind the indicator lights, a bulletproof shield that emerged over the back window, revolving license plates valid in Britain, France, and Switzerland, a high-pressure oil jet, a device for dispensing nails, a rear smoke screen, and a revolving tire slasher that emerged from the hubcaps. As if that wasn't enough, it even had an ejection seat that removed unwanted passengers, an onboard radar unit for tracking vehicles, and a weapons tray beneath the driver's seat.

The car itself was also a powerhouse; it had a 4-liter engine that produced 282 hp at 5,750 rpm, a five-speed transmission, and a top speed of 150 mph. Bond used the radar unit in the car to track Goldfinger to Switzerland. It was activated by a homing device that Bond put in Goldfinger's car immediately after he played golf with him. The weapons in the car were used mainly in the chase scenes near Goldfinger's factory. And I'm sure everyone was waiting for him to use the ejection seat—as I was. He used it on one of Goldfinger's guards, who was taking him to the factory. It's perhaps a little strange that they allowed Bond to drive his own car to the factory after he was captured, particularly after he had already used some of his "tricks" on them. Goldfinger was quite amused by the car.

The Aston Martin DB5 was also used in *Thunderball*. After catapulting over the wall using a jet pack in the opening scene, Bond escaped in the DB5. He was soon being chased, and it wasn't long before a new weapon came into play. He shot a hefty stream of water at his pursuers with his new "water canon." I have a problem with this, though. The pressure and amount of water that came out of the cannon was a little too much for such a small vehicle. It would have to have had a huge storage tank and pressure unit stored somewhere, and there certainly wasn't enough room in the DB5.

The DB5 also made an appearance in *GoldenEye* in 1995 and in *Tomorrow Never Dies* in 1997. Different cars were used, however. The original DB5 that appeared in *Goldfinger* was bought by a collector in 1986 and for several years it was exhibited at exhibitions and displays. In June 1997, however, it was stolen and it has never been recovered.

Toyota 2000 GT

For the fifth Bond film, *You Only Live Twice*, Bond was in Japan so it was only natural that his car be a Japanese make. I have to be careful in calling it "his" car, however, because he never actually drove it. It was driven by his Japanese accomplice, Aki. The car, a Toyota 2000 GT, was used mainly because it had just caused a sensation at the Tokyo Motor Show. Among its more interesting features were a closed-circuit TV system, a HiFi receiver and cassette player, a two-way radio, a voice-controlled tape recorder, and a miniature color television, all of which were quite new at the time. It had an inclined six-cylinder engine with a horsepower of 150 at 6,600 rpm, a five-speed transmission, and top speed of 143 mph. It also featured independent suspensions in the front and back and disk brakes in the front and back, which were well ahead of their time.

Wet Nellie

The Aston Martin was Sean Connery's car, although in later years it was also used extensively by Pierce Brosnan. And just as Connery had his car, Roger Moore also had his own car—the Lotus Esprit. It was, in fact, no accident that the Lotus Esprit was used in several Bond movies. The Lotus manager, Don McLaughlin, heard that Pinewood Studios in London were making preparations for a new Bond movie to be titled *The Spy Who Loved Me*, and he parked one of the new Lotus cars directly in front of the studios. It worked. Producer Albert Broccoli spotted it and decided to use it in the movie.

The thing that made the Lotus Esprit such a hit in *The Spy Who Loved Me* was its ability to transform itself into a submarine. This wasn't something you saw every day. Aside from this, though, the Lotus was also well equipped with weapons; it had torpedoes, a surface-to-air missile, depth charges, and a smoke screen for both land and sea. Furthermore, it had a 140-hp engine, a five-speed transmission, and a maximum speed of 135 mph; it could accelerate from 0 to 60 in 9.2 seconds.

The most dramatic scene involving the Lotus begins with Bond and Amasova being chased through the mountains by several of Stromberg's assassins. Stromberg obviously didn't have too much faith in them, though, as he also sent his assistant Naomi in with a helicopter to make sure the job was done properly. I particularly like the part where she winks at Bond, just before she tries to do him in. The chase ends when the Lotus plunges off a jetty into the sea. It looks like the end for Bond and Amasova, but the Lotus quickly transforms itself into a submarine. Then, unexpectedly, it blasts off a missile that makes short work of Naomi and her helicopter. The "wink" obviously didn't do her much good. All in all, it was quite a scene.

Although some trick photography was involved, the car actually did function as a submarine, and it had a fairly respectable speed. The underwater transformation from a car to a submarine wasn't easy, though; it involved five different cars, each one representing part of the transformation. First the wheels came up and the wheel arches were filled in, then fins appeared and the back bumper slid up and propellers appeared. Quite a feat!

The "Lotus submarine" then made its way to Stromberg's Atlantis headquarters where it was attacked by a number of frogmen using motorized underwater sleds. But Bond used the Lotus' weapons and prevailed. In the final scene the Lotus drove out of the water and onto the beach, much to the surprise of the sunbathers. Again, a little ingenuity was used. The vehicle was actually pulled out of the water on hidden rails, but it all looked quite realistic. Incidentally, the Lotus was referred to as "Wet Nellie," after the autogyro, Little Nellie, and over the years it has become the second most famous Bond vehicle. Where is it today? It's owned by the Ian Fleming Foundation.

A Lotus Esprit was also used in 1981 in *For Your Eyes Only*. The new Lotus was a Turbo; two Turbo Lotuses were actually used in the film. Gonzales's men attempted to break into the first one; they used a sledge hammer on it, but one of the car's weapons came into play—a self-destruct unit—much to their surprise (and demise) the car blew up in their faces. Bond appeared later with another Lotus, however; both cars had a engine capacity of 2.17 liters, a horsepower of 205 at 6,000 rpm, and a top speed of 148 mph; the cars could go from 0 to 60 in 6.6 seconds.

The Most Unlikely Bond Car

The Lotus played a central role in *For Your Eyes Only*, but in many ways it was overshadowed by what can only be thought of as "the most unlikely Bond car." It certainly wasn't a sports car, but it was a popular car in France, where it had sold millions and had developed a cult following. It was the boxy Citroën 2VC. One of the more exciting chase scenes in the Bond movies involved this car. It was obvious to the producers that the standard stock Citroën wasn't up to the demands that were going to be made on it during the chase scene, so they modi-fied it. Its power was increased from 29 hp to 59 hp, which is still pretty low by Bond standards. Nevertheless, it was capable of a top speed of 100 mph. Its chassis was also lengthened slightly, the transmission and clutch housings were modified, special shock absorbers were installed, along with heavy-duty stabilizers and a roll bar. And I have to say that it did the job well.

More Power, More Speed

With a new Bond—Timothy Dalton—in *The Living Daylights* came a new Bond car, an Aston Martin V8 Volante. Bond used it to get out of communist Czechoslovakia while being chased by the Czech police and the KGB. The most powerful Bond car to that time, it had a 5.34-liter engine and 300 hp, which was considerable for 1987. Its maximum speed was 146 mph and it went from 0 to 60 in 6.6 seconds. Q "winterized" it for all the snow and ice in Czechoslovakia by adding retractable outrigger skis for stability on the ice and snow and spikes that emerged from its tires to give it a better grip on ice. And finally, it had an ice-cutting device, which helped get rid of a pursuing car on a frozen lake.

Most amazing were its array of weapons. For extra thrust, it had a jet that was used in a jump over a dam. Behind the front fog lights it had guided missiles that were guided to their target by using a display on the inside of the front windshield. And in the front hubcap was a laser that was used to cut the chassis off a police vehicle that pulled up beside them. (This was one of the few comical scenes in the movie.) All in all, the Aston Martin was quite a vehicle, but in the end it blew up. Like several of Bond's other vehicles it had a self-destruct mechanism just in case someone tried to get into it.

BMW Z3

While preparations were being made to film *GoldenEye* the producers decided to use the BMW Z3 as Bond's official car. There was a new Bond, namely Pierce Brosnan, and it seemed appropriate that he be seen in a new car. As it turned out, there was more publicity about the new car than there was use of it in the film. A press conference was even held to announce the use of the Z3. Forty TV stations from around the world and about 500 reporters attended, but strangely, the car was not shown. Everything was kept secret. Finally, in November 1995, when *GoldenEye* premiered in New York, the Z3 was presented to the press in Central Park by Pierce Brosnan and Q. Twenty Z3s were available and many of them were sold, some to well-known actors, including Madonna, Alec Baldwin, and the director, Steven Spielberg.

The Z3 had a five-cylinder engine with 140 hp and a top speed of 127 mph. It didn't have as many extras as some of the previous Bond cars, but it did have a parachute-braking system, Stinger missiles, and an all-points radar system.

In the movie, the Z3 was actually overshadowed by the Aston Martin DB5. Bond was driving it near Monte Carlo when Xenia Onatopp pulled up beside him in a Ferrari 355 GTS, and soon the race was on. The Ferrari was on loan, and during the race it suffered considerable damage. Ferrari decided they would not ask EON Productions to pay for the damages as long as the Ferrari won the race. So, indeed, Bond lost the race—all for a good cause.

Bigger Is Better

Gone were sports cars in *Tomorrow Never Dies*. Bond was now driving a BMW 750iL sedan, which had to be something new for him. But like his previous cars it was well equipped, and, interestingly, it could be driven by remote control using a pad hidden in a mobile phone. The phone had a video screen that showed the view through the front window of the vehicle. With a 5.4-liter engine that gave a top speed of 155 mph, it was the most powerful Bond car to that time.

Mounted in the sunroof were 12 heat-seeking missiles, and it had an antitheft system that immobilized the vehicle and delivered a powerful shock to anyone trying to break into it. Below the logo in the front was a chain cutter that came in handy when a cable was strung up in front of it during one of the chase scenes. In addition it had smoke and tear-gas discharge units, reinflatable tires, and bulletproof glass and body. And finally there were spikes in the trunk that could be used against anyone chasing the car, and they were used to good effect.

The Ultimate Driving Machine

In *The World Is Not Enough* Bond graduated to a BMW Z8, which was even more powerful than the 750iL. Furthermore, he was back in a sports car, which seemed a little more natural for him. It had a 5-liter V8 engine and 400 hp, with a maximum speed of 155 mph, and it could accelerate from 0 to 60 in 4.4 seconds.

The Z8 had a bulletproof windshield and armor-plating protecting the driver and passenger, and it was armed with two Stinger missiles. It also had a navigation system and a laser for listening in on conversations in nearby cars or buildings. One of its most interesting features, however, was that it could be started remotely using a device hidden inside the ignition key, and Bond could direct it to pick him up. In the movie, a gigantic saw blade suspended from a helicopter sawed it in half. Such a waste of a beautiful car.

Bond used one of his missiles on one of the two attacking helicopters, so he had at least some revenge.

Die Another Day Vehicles

Three different vehicles were used in *Die Another Day* and, as you might expect, Bond's vehicle was even more powerful than his previous ones. After dabbling with BMWs for a couple of movies he was back with an Aston Martin. This time it was a 2002 V12 Vanquish with a 5.9-liter engine and 460 hp at 6,500 rpm. Its top speed was 190 mph, and it went from 0 to 60 in 4.6 seconds.

The weaponry included missiles behind the front grille, auto-targeting machine guns under the air take in the hook, and machine guns in the front. Like the DB5 in *Goldfinger* it had an ejection seat, and for ice, it had tires with retractable spikes. Something interesting was a new feature, namely, invisibility; I'm not sure how they would manage this in the "real world"—it would certainly be a neat trick.

Bond wasn't the only one with a fancy weapon-laden car. The

villain, Zao, had a Jaguar XKR and it was also well equipped. It has eight cylinders and a 3.9-liter engine with 370 hp at 6,150 rpm, and it had a top speed of 155 mph. Its weaponry included a Gatling gun, missiles, rockets, a spear, and mortars in the trunk. One of the major action scenes in the film was a battle between the two cars on a frozen lake.

A Ford Thunderbird also made an appearance in the film. Jinx arrived at the Ice Palace in Iceland in a Thunderbird. It didn't have any extra equipment, but it was equipped with a 3.9-liter engine that produced 252 hp at 6,100 rpm.

Other Cars

For completeness I'll briefly talk about some of the other cars used in the Bond movies. None of them played a central role, but they were interesting nonetheless. Everyone remembers Goldfinger's black and yellow Rolls Royce that he used to smuggle gold to other countries by melting the fenders down. It was a 1937 model with a 7.3-liter V12 engine, so it was no slouch in the power department. Bond followed it to Switzerland and watched Goldfinger's mechanics dismantle it.

The Man with the Golden Gun had two particularly interesting cars. In one of them, the car itself wasn't particularly memorable, but the fact that it turned into an airplane was. Bond chased Scaramanga's AMC into a large shed, and when it came out the other end it had wings. He flew off to his private island leaving Bond a little speechless. But, of course, Bond soon found out where he was headed and followed him in a seaplane.

The other car in *The Man with the Golden Gun* was also an AMC, and like Scaramanga's car, it wasn't much to write home about, but it performed one of the most amazing tricks in the Bond movies. It did a 360-degree barrel roll. The only problem was that it happened so fast you barely had time to see it before it was over.

I'm a little disappointed that one of my favorite cars barely makes an appearance in the Bond movies. I'm talking about the Corvette. It does, however, make a brief appearance in A *View to a Kill*. After listening in on Zorin's pipeline in San Francisco, Bond bumps into KGB agent Pola Ivanova and they go to a Chinese spa house in the Corvette. This Corvette had a V8, 5.7-liter engine with 205 hp at 4,300 rpm and a maximum speed of 137 mph; it could do 0 to 60 in 7.1 seconds.

Even the lowly Volkswagen Bug makes a brief appearance. Yes, I know . . . there are a lot of Volkswagen fans out there, so don't get me wrong. I'm not degrading it in any way. It's a great car. The one that appeared in the Bond movies was in *Octopussy*. Bond got a lift from a German couple in a Bug. In case you're interested, it had a 1.6-liter engine and 50 hp, with maximum speed of 80 mph. It could hardly compete with the sports cars Bond drove, but it's an elegant car in its own way, nevertheless.

Finally, I should mention that Jinx's Thunderbird wasn't the only one used in the series. A 1970 Thunderbird appeared in *Diamonds Are Forever*. The villains Wint and Kidd drove it. It was a V8 with a 7-liter engine and 51 hp.

Other Vehicles

Cars were not the only type of vehicle used in the Bond movies, of course. Motorcycles were used in many of the films. One of the most memorable was the BMW R1299 used in *Tomorrow Never Dies*. Bond and Wai Lin sped through the streets of Saigon on it. They were handcuffed together at the time, but Wai Lin worked the clutch and Bond steered and accelerated. The most exciting part of the chase was the jump over the helicopter.

A BSA 650cc Lightning motorcycle was used in *Thunderball* by Fiona, the head of SPECTRE's execution branch. It was equipped with rockets and the sequence where they were fired almost turned into a disaster. The rockets, which were real, were fired from the motorcycle at a car driven by stuntman Bob Simons. For effect the car had its trunk filled with explosives. On the first take things went wrong when the rocket penetrated the armor set up to protect Simons. He had to make a quick exit to save his life. Luckily, everything went okay on the second try. Motorcycles were also used in *For Your Eyes Only* and *GoldenEye*.

Closely associated with motorcycles is the "moon buggy" used in *Diamonds Are Forever*. In the desert near Las Vegas Bond escapes from Whyte's guards and, while running through an astronaut-training ground, he commandeers a moon buggy. He is chased by a squad of cars, but they are not as efficient on the sand dunes as the moon buggy and he soon loses them.

Perhaps the most exotic "other vehicles" are the huge trucks used in *Licence to Kill*. They are involved in some exciting scenes. Sanchez sets out from his laboratory with his fleet of Kenworth tankers, and Bond is determined to stop him. The scenes that followed involved a lot of action as several of the tankers crashed.

And, finally, one of the strangest vehicles that Bond drove — in St. Petersburg he "borrowed" a tank from the Russians and pursued the villain's car through the streets of the city. I'm not sure where he learned to drive a tank, but he managed.

The Race Is On

Car Chases

With fast cars come car chases, and there are plenty in the Bond films. Almost every film has at least one, and they are exciting.

As the car swerves, bounces around, and tumbles, the driver can go all the way from weightlessness when he is in midair, to double or triple his usual weight when the car hits the ground. Not only are there forces in the upward and downward directions, but severe forces are exerted on the driver when a car goes around a corner. They are referred to as centripetal forces. And of course when a driver brakes or accelerates he also generate tremendous forces, assuming he's strapped in. (If he's not, things can be a lot worse.)

There are a lot of thrilling car chases in the Bond films and it's hard to choose between them. Each is a little different in its own way. One I liked, in particular, was the Citroën chase in *For Your Eyes Only*. It was quite unusual, and involved a very "different" car. And it was fun to watch.

I can't say I've ever experienced the excitement (or scare) of a chase like the chases in the Bond movies, but there is one ride I remember quite vividly. When I was in high school I caught a ride one day with a fellow named Hank. I didn't know Hank, and when I looked at his beat-up half-ton truck I was a little apprehensive. Nevertheless, I jumped in the passenger seat beside him. I was barely settled when he took off, and when I say "took off" that's what I mean. The road was bumpy and he literally flew over the bumps. I was used to fast driving, but this was out of my league.

I'm sure my face was white and he must have noticed, because he finally looked over at me and said, "Don't worry, I've driven this road dozens of times."

I was in too much shock to reply. But what really had me worried was that the road zigzagged along what appeared to be the edge of a cliff. I kept looking at it out my side window, and as far as I was concerned we were too close to it for comfort.

Suddenly Hank swerved his beat-up truck to the right. I couldn't believe it! Sure that we were headed over the cliff, I closed my eyes and covered them. I could feel my stomach rise into my throat. But strangely we didn't hit; we banged around a lot but when I finally got up enough nerve to open my eyes I saw that we were on the level again.

Hank laughed as he looked over at me. "I'll bet that surprised you," he said nonchalantly. I was in too much shock to say anything. "It was a shortcut," he said.

Whenever I watch the Citroën race, particularly when Bond leaves the road and heads "cross country," I think of my ride with Hank.

The Physics of Car Chases

Anyone in a car chase experiences a lot of different forces, and of course, *force* is one of the major things we study in physics. Force comes about when we accelerate, and as we saw earlier, acceleration has units of ft/sec^2. I'm sure, however, that you've also heard of it referred to in terms of "g's." Astronauts and even race drivers are quite conscious of how many *g*'s they're experiencing at any time. They have to be, because they know that if they experience too many, they'll black out, and that's not in their best interest.

One *g* is 32 ft/sec^2, and with an acceleration or deceleration of 1*g* you will experience a force on your body equal to your body weight. If you weigh 160 lbs, 2*g*'s will be 320 lbs, 3*g*'s will be 480 lbs, and so on. And as I'm sure you know, 320 lbs is a pretty hefty weight to try lifting.

What is the maximum number of *g*'s a human can experience without blacking out? The answer is about 9, and you can take this force only for a few seconds. Jet pilots occasionally encounter this many *g*'s when they go into a tight turn, and they know they have to be careful.

One place where people experience exceedingly high *g*'s is in car crashes. The force is so severe that most people do not survive crashes that involve high *g*'s. The highest *g* that a human has ever been known to survive is 86*g*, and that was only for a fraction of a second. Most of the time when we accelerate or decelerate we experience relatively mod-est *g*'s—thank heavens. In a commercial airliner we encounter about 1.5*g*'s, and even in the space shuttle the takeoff is at about 3*g*'s.

What we are interested in is how many *g*'s we experience in going around a curve in a car. Everyone knows that there is an outward force when they go around a curve, and most people can identify it as either a centripetal or centrifugal force. But which of these two is it? Because some confusion exists about these terms let's spend a few mo-ments discussing them.

We'll begin with a simple example: a ball being whirled around on the end of a string (fig. 42). Two forces appear to be involved: an out-ward force pulling on the string, and on your arm (assuming you are the one whirling it), and an inward force pulling on the ball on the end, forcing it to go in a circle. The outward force is usually referred to as a centrifugal force, and the inward one as a centripetal force. They appear to be equal and opposite, and this seems to make sense since Newton's third law tells us for every action force there is an equal and opposite re-action force. But if the centrifugal force was acting directly outward, and

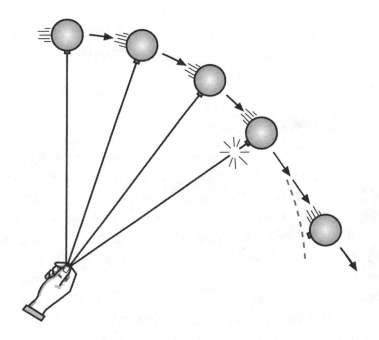

Fig. 42. A ball being whirled at the end of a string. There is an inward centripetal force.

you cut the string, the ball should fly off "in a outward direction." And it doesn't; it flies off tangential to the circle it is tracing out. Something appears to be wrong. Let's look at this a little closer. First, we know that Newton's third law applies only to systems at rest, and the system above is not at rest. Second, if the two forces actually were equal and opposite, they would cancel and we would have no motion—but we do have motion. These problems (and others) tell us that the outward "centrifugal force" is actually a "fictitious force." In other words, it doesn't really exist.

There is a case, however, where the centrifugal force has significance. In the case of the ball and string we are "outside" observers; in essence, we are observing the system from the outside. In this case the centrifugal force is fictitious, but if we could jump on the ball and ride along with it, the centrifugal force would have meaning.

Well, I guess that's enough about that. Let's get on with centripetal forces in cars. We know that the acceleration (a) when we round a curve of radius R at velocity v is given by

$$a = v^2/R.$$

We can convert this to g's by dividing by 32; also if we're dealing with miles per hour rather than feet per second we need a conversion factor of 22/15. Let's assume we are going around a curve of radius 100 feet with a speed of 60 mph (88 ft/sec). Substituting in we get an acceleration of 2.42g's, which is fairly large. At a radius of 50 feet this would be 4.84g's, which would be extremely uncomfortable. But there's something we haven't taken into account so far, the traction of the tires on the car. How many g's can they take before they slide? As it turns out, they can take only about 1g, so it's not much use talking about high g's in the case of a car rounding a curve. It will slide well before the g's get very high.

Sliding depends, of course, on the traction of the tire, so let's look at it a little more closely. Traction is a measure of how well the tire sticks to the road, and this depends on what is called the *contact patch*, an oval region of the tire that is in contact with the road (fig. 43). The position of the contact patch changes, but it is always roughly the same size and shape.

How do we determine when the tire slips? We use the formula

$$F \leq \mu W$$

where F is the force on the tire, W is the weight acting down on it, and μ is the coefficient of friction. Slipping occurs when the inequality sign ($\leq$) becomes an equal sign. In other words, when

$$F = \mu W,$$

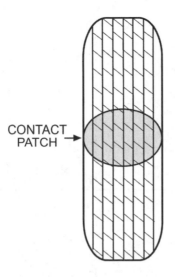

CONTACT
PATCH →

Fig. 43. The contact patch on a tire

but we know that $F = ma = \mu mg$, where W is equal to mg. Canceling the m's gives us the acceleration just as slipping occurs, namely,

$$a = \mu.$$

Note that it is independent of the weight on the tire, which may seem surprising. In practice, other factors come into it, and the weight does have a slight effect.

The coefficient of friction, μ, ranges between 0 and 1, where anything sliding on ice is close to zero, and a good tire on a dry road is close to .9. Since .9 is nearly one, we'll assume that, to a good approximation, the tire will slide at a g of about 1. Let's use this to describe what is called the *traction circle*.

We begin by drawing a circle as follows, where there is slippage outside the circle and traction inside (fig. 44). The arrow represents our acceleration. Note that we can accelerate or decelerate (brake) up to 1g

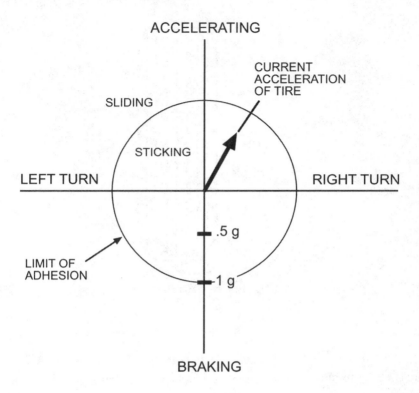

Fig. 44. The traction circle

without slipping as long as we don't turn. But the moment we turn at this acceleration we will slide. Similarly, we can corner to the right or left up to 1g, as long as we don't accelerate or decelerate.

Sliding can be dangerous, because we may lose control of the car, but a small amount can actually be advantageous, since more tread surface is coming into contact with the road and the effective contact patch is therefore larger.

Braking, accelerating, and cornering have another effect on the car that is important in car chases: they redistribute the weight of the car. In other words, they shift it around. Assume that initially (at rest) the car has its weight uniformly distributed over the front and rear axes. In practice this is not necessarily true, but we will assume it for

DEATH RAYS, JET PACKS, STUNTS & SUPERCARS

simplicity. If you accelerate, weight will be shifted to the rear of the car, so more than half the weight will now be on the back axis. As an example, assume the car weighs 3,000 lbs; at rest, 1,500 lbs will therefore be on the front axis, and 1,500 lbs on the back axis. With an acceleration you could shift 500 lbs to the back axis. This would mean that 2,000 lbs is now on the back axis and 1,000 lbs is on the front one.

What happens when this occurs? Mainly, the steering changes. With less weight on the front axis, you're likely to understeer. Also, with more weight on the back axis, the back brakes will be slightly more effective.

You get a similar effect when you brake. In this case weight is shifted forward to the front axis. You could shift 500 lbs forward so that 2,000 lbs is now on the front axis, and 1,000 on the back. And again the steering will be seriously affected (you will tend to oversteer), and braking may be slightly affected.

You get a similar effect when you swerve to the right or left. In this case you are dealing with the centripetal force. One of the things you have to worry about in turning rapidly is upsetting the car—something you frequently see in car chases. When a car goes around a corner the centripetal force on it tends to make it roll. The "roll force" is countered by a frictional force between the road and the tires. The car will lean, but if it has a good suspension system, the lean will be kept to a minimum.

If the roll force is great enough, however, the car can topple. Let's consider this. If a car rolls, it obviously must roll around an axis, which we refer to as the *roll axis*. The position of this axis can be calculated relatively easily, but I won't go into detail here (the details can be found in my book *The Isaac Newton School of Driving*). In short, you use the suspension system to determine the *roll center* of the front and back axes, then you draw a line between them to get the roll axis (fig. 45). This is the axis the car rolls around when it rounds a corner.

For roll to occur, we need a torque or twisting force. Earlier we saw that torque is defined as a force times a lever arm. Where is the lever arm in this case? To answer this we have to consider what is called

There is some interesting sliding on the corners, but not a lot of physics is involved.

A Chase on Ice

A chase of another type comes in *Die Another Day*. It's a one-on-one race between the villain Zao and Bond, but what is different is that Zao's car is almost as well equipped as Bond's. And, the chase takes place on ice, which brings back memories for me. I've had several narrow escapes on icy roads when my car has begun to slip, and I can assure you they were scary. Traveling down the road at 40 or 50 mph backward is not my idea of fun, but I have done it a couple of times.

Bond's car is invisible during the first part of the race (I'm still not sure how they could really manage this), but it doesn't help him. Zao has an infrared detector and can see its heat. At one point Zao shoots a missile at Bond's car and knocks it upside down. It slides for some distance on its roof, but this doesn't appear to perturb Bond. He uses the escape hatch to knock it upright, and soon he is chasing Zao again.

Bond has more on his mind than the car chase. He knows that Jinx (the heroine) is in danger and he breaks off the chase and heads to the ice palace to save her. Some of the more spectacular shots come after he crashes through the icy walls of the palace. Zao is still in pursuit of him and Bond zigzags through the narrow corridors of the palace, and in the end he rescues Jinx and does away with Zao.

There are many other amazing chases in the films, of course. Two worth mentioning are the motorcycle chase in *Tomorrow Never Dies* and the tank chase in *GoldenEye*. Both are quite spectacular.

Bond in Space

tar Wars was released in 1977. *The Spy Who Loved Me* was released at about the same time. Both were a success, but *Star Wars* was a much bigger hit, and this wasn't lost on Albert Broccoli (producer of the Bond films). He was awed by the success of *Star Wars* and decided to do something about it. He had planned on bringing out the film version of Ian Fleming's book *For Your Eyes Only,* but it was quickly put on hold. Fleming had also written a book about space, titled, *Moonraker,* and although it was dated, Broccoli was sure it could be brought up to date. (Aside from the first few books they never followed the books, anyway.) In the end there was little of the book in the film. Fleming's story was tame compared with *Star Wars.* But if you look carefully at the film version of *Moonraker* you see that something else is a bit of a problem. The plot is almost exactly the same as that of *The Spy Who Loved Me.* They both have a megalomaniac who wants to take over the world and change it. Despite this, *Moonraker* was a tremendously successful film. The critics didn't like it, but the public loved it.

The space race had been around for many years when *Star Wars* and *Moonraker* came out; I remember it quite vividly. Like many teenagers in the 1950s I was enthusiastic about rockets and the possibility of space satellites (I suppose that's why I went into physics), but when the first satellite was put in orbit in 1957 I think I was as shocked as everyone else. It was hard to believe that a small satellite was actually in orbit around the Earth. Part of the shock was, of course, that the satellite wasn't American; it had been put in space by the Russians. The satellite was called "Sputnik" and it was soon a word that was etched in the minds of Americans everywhere. It was a small satellite with a diameter of only 23 inches and a weight of 184 lbs, but it was still an *orbiting* satellite. I remember that shortly after it was launched my "least favorite" professor announced to the press that he had calculated its orbit. "Wow!" I thought. "Maybe he wasn't that bad after all." As we will see, though, it isn't too difficult to calculate an approximate orbit.

How did the Russians manage to put a satellite in orbit ahead of us? It was hard to accept, and needless to say, the government was in panic. We obviously had to do something—fast! And the announcement came almost immediately: we would be launching a satellite into space in December 1957. But as everything was being readied for our first launch we got another shock: the Russians put a second satellite into orbit, and this one wasn't tiny—it weighed 1,118 lbs and aboard it was the first cosmonaut. True, it wasn't a man; it was a dog; nevertheless, it was still a cosmonaut. The dog's name was Laika.

In December we were ready with the Vanguard rocket. It was to be launched from Cape Canaveral, Florida, and reporters from around the world assembled to watch it. The countdown began on December 5. When it reached zero, the engines blasted and the Vanguard rocket began to rise. Then—to everyone's dismay—it fell back on the pad and exploded in a brilliant ball of orange. It was a monumental disaster— and an embarrassment.

Fortunately, there was a backup. Wernher Von Braun, who had built the German V-2 rocket a few years earlier, and his group had been working on a Jupiter-C rocket, and in less than 60 days it was equipped with a tiny satellite called *Explorer I* and was ready to go. On January 31, 1958, the countdown began and at 10:48 p.m. the rocket rose from its pad and disappeared into the night sky. Americans sighed in relief. The launch was successful.

Space Physics

Putting a satellite into orbit involves a lot of complex physics, and things are even more complicated if the rocket is going to the moon or one of the planets. Everything depends on the gravitational fields of the objects involved. How do we deal with these fields? We obviously have to know a lot about them, and, as it turns out, there are two theories of gravity: Isaac Newton's and Albert Einstein's. I'm sure you've heard of both of them. Newton's theory is the older of the two by about three hundred years. It tells us that every massive object in the universe attracts every other massive object with a force that is proportional to the product of their masses, and inversely proportional to the distance between them squared. I'll admit, that's quite a mouthful. It scared almost everybody when Newton first wrote it down, but they soon realized it was a stroke of genius. Besides, it's not really as bad as it looks. In terms of mathematical symbols we can write it as

$$F = GMm/R^2$$

where G is the gravitational constant, a constant that governs the gravitational field throughout the universe, M and m are the masses involved, and R is the distance between the masses. From this you can see the product of the masses, and also the dividing factor (inversely proportional to) R^2.

Einstein's gravitational theory is much more recent than Newton's, of course. Newton saw gravity as a mysterious "action-at-a-distance" force, something that Einstein didn't like and replaced with a "curvature" of space. For most people this curvature was even stranger and more mysterious than an action-at-a-distance force. Could you see it? What was it? No, you can't see it or even visualize it properly. But you can describe it using mathematical equations, and that's all that really matters.

Einstein's theory turned out to be much more accurate than Newton's, but this doesn't mean that Newton's theory is old fashioned and crude; it's actually a highly accurate theory. Indeed, it is accurate enough that Einstein's theory is not needed when making calculations of orbits and space flight. What is Einstein's theory good for, then? Scientists use it to study the overall universe and exotic objects such as black holes; it's much better than Newton's theory in these cases. For orbital calculations, though, Newton's theory is good enough, but don't get me wrong: the calculations are still so complicated that computers are needed. In fact, it can be safely said that without computers space travel of any kind would be impossible.

Aside from Newton's law the most important laws related to orbits in space are Kepler's laws. Johannes Kepler, a seventeenth-century German scientist, formulated three laws that are named after him. He used them to describe the orbits of the planets, but they also apply to the orbits of rockets and satellites in space.

KEPLER'S FIRST LAW: **Each planet moves in an elliptical orbit with the sun at one focus.**

An ellipse is an egg-shaped curve that can be very elongated or circular (a circle is a special case of an ellipse; for more details see below). The Earth, for example, orbits the sun in an ellipse, and the sun is at a special point within the ellipse called its focus. What is a focus? The best way to understand it is to look at how ellipses are drawn. You take a length of string and tack its two ends down so that the string between them forms a loose loop. Using a pencil, pull the loop tight and draw the resulting curve; it will be an ellipse and the two tacks are at its foci (fig. 47).

KEPLER'S SECOND LAW: **The line connecting the sun and a planet sweeps out an equal area in equal times.**

DEATH RAYS, JET PACKS, STUNTS & SUPERCARS

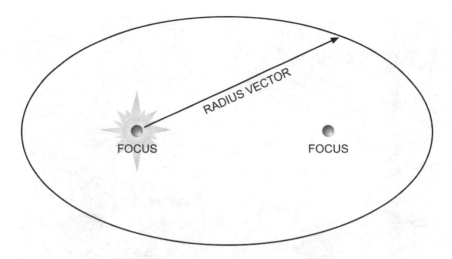

Fig. 47. An ellipse showing the two foci. The sun is at one focus. The distance from a focus to the ellipse is called the radius vector.

Looking at the diagram, we see one of the major consequences of this, namely, that if equal areas are to be swept out in time the planet will travel faster when it is closer to the sun (fig. 48). And indeed, planets do not travel at the same speed in their orbit; their speed varies.

> KEPLER'S THIRD LAW: The square of the planet's orbital period (time to go around the sun) is proportional to its average distance from the sun cubed (P^2 is proportional to R^3).

These three laws are particularly helpful in determining orbits in space. When a rocket takes off from Earth it can go into one of four possible types of orbits: circles, ellipses, parabolas, and hyperbolas. I'm sure you've heard of circles and ellipses, but parabolas and hyperbolas may be new to you. The best way to visualize these curves is to imagine slices across a cone as shown in the diagram. If you cut across parallel to the base (a) you get a circle. If you cut across at an angle (b) you get an ellipse, and it's easy to see that you get ellipses of different elongations,

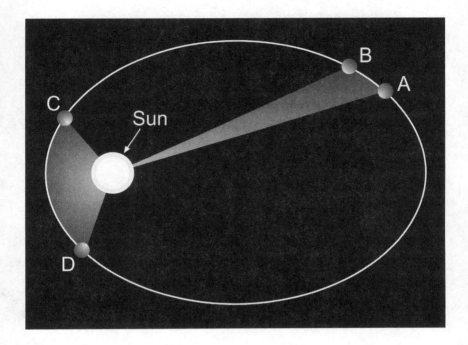

Fig. 48. A demonstration of Kepler's second law. Note that the planet travels further in the same time when it is closer to the sun

depending on where you slice the cone. If the cut intersects the base (c) you get a parabola, and finally if you cut perpendicular to the base (d) you get a hyperbola (fig. 49). The first two of these are closed orbits, and rockets sent into these types of orbits will orbit the Earth; the second two are open orbits and the rocket will escape to space in this case. In most cases, assuming we are below the escape velocity (see the next section) the orbit will be an ellipse.

Escape Velocity

The main thing you have to do to place a satellite into orbit is overcome the gravitational field of the Earth. You know if you throw a rock upward, it rises to a certain point and falls back to the ground. If

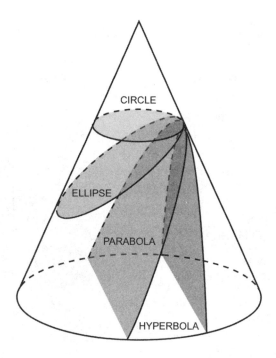

Fig. 49. Slices through a cone showing the various curves

you could throw it with a high enough speed, though, it would completely overcome the gravitational pull of the Earth and escape to space (fig. 50). This velocity is referred to as the *escape velocity*. The velocity needed to put a satellite into orbit, however, is less than this, since it has not completely escaped the gravitational pull of the Earth.

How do we determine the escape velocity? Let's begin by considering the *energy* of an object in space. Two types of energy are of interest to us: potential energy (PE) and kinetic energy (KE). The potential energy of a small object at a distance r from a large body such as the Earth is defined as the work that has to be done by an external agent to bring this body from a very great distance (essentially infinity) to a distance r from the large body. We write this mathematically as

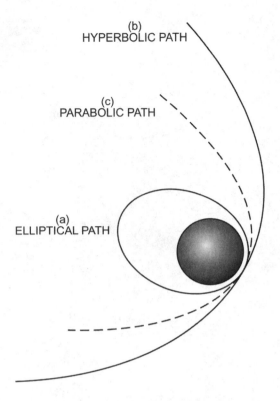

Fig. 50. Various orbits around the Earth. Outer ones are open, the inner ellipse is a closed orbit.

$$PE = -GMm/r,$$

where m is the mass of the small body, M is the mass of the large body, and G is the gravitational constant we discussed earlier. (Don't worry about the minus sign; it's merely convention.)

One of the first things we learn about energy is that it is conserved. This means it can't be created or destroyed, only changed in form. In regard to space flight we are concerned only with potential and kinetic energy. We looked at kinetic energy earlier and saw that it is energy of motion with the formula $\frac{1}{2}mv^2$, where v is the velocity of the object. If the total energy is to be conserved it must remain constant, and this means that KE + PE = constant. Substituting the formulas above into this, we get, after canceling the m values,

$$v^2 = 2GM/r.$$

This is the escape velocity, which is also sometimes referred to as the parabolic velocity because it is the velocity that gives a parabolic or open orbit. Substituting numbers into it for the escape velocity of Earth we get 25,000 mph. This means we need a speed at takeoff of 25,000 mph to completely escape the Earth's gravitational pull. This is the speed at which we would also have to throw the rock in the earlier example; it's obviously far beyond our ability. In most cases, however, we merely want to put a rocket or satellite into orbit, so we don't need this speed.

Let's look at the launching of a rocket in more detail. In most cases it is difficult for a single rocket to provide the velocity needed to put a satellite in orbit. This is overcome by what is called *staging*; by this I mean that a three-stage rocket is used with the first stage on the bottom, the second stage directly above it, and the last stage at the top. The third stage determines the final orbit of the satellite; in fact, it is part of the final orbit, as shown in figure 51.

In most cases the orbit will be an ellipse. What is particularly desirable, however, is an orbit that is as circular as possible. The reason is that, in an ellipse, part of the orbit will be much closer to the Earth, and if this section lies in the outer region of the Earth's atmosphere (as it sometimes does), when the satellite passes through it, there will be considerable drag on it. This drag tends to change the orbit and will eventually bring the satellite down. If the orbit is circular there is little drag (or at least it is uniform).

In addition to circularity, several other properties, or types of orbits, are desirable. One is referred to as a *geosynchronous* orbit; it has

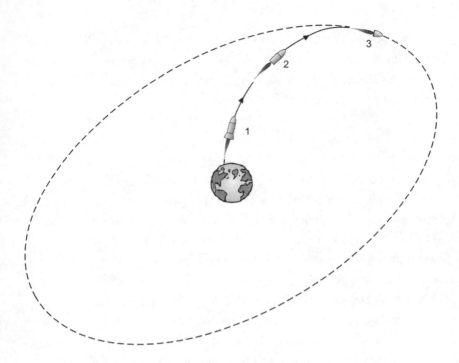

Fig. 51. A three-stage rocket. The final stage determines the orbit.

an orbital period that is exactly one day long and is therefore close to the rotational period of the Earth. Communication satellites are usually put in geosynchronous orbits. Even more desirable are what are called *geostationary* orbits; they are geosynchronous orbits that are over the equator. In this case the satellite would appear stationary in the sky; again, orbits such as this are desirable for telecommunication and weather satellites.

Maintaining such an orbit, however, is easier said than done. The problem is perturbations, or small forces that pull the satellite out of orbit. One of the major perturbations is caused because the Earth is not a perfect sphere; it is spinning, and therefore it bulges slightly at the equator. The sun and the moon cause other perturbations; they also attract the satellite, and their attraction is greater when the satellite is in

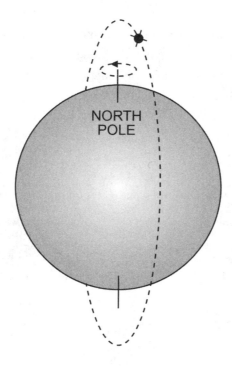

Fig. 52. A polar orbit

their direction. There are points in the Earth-moon system (also in the sun-Earth system), however, at which the gravitational pull is balanced. In other words, the pull of the Earth is equal to the pull of the moon. These are referred to as Lagrange points, and in certain cases satellites are placed at these points.

Another desirable orbit is referred to as a *polar orbit* (fig. 52). In a polar orbit, the satellite moves around the Earth in a direction perpendicular to its spin, so that it passes through the north and south poles. With the Earth spinning under it, this orbit gives a maximum visibility and is ideal for extended observations. The entire surface of the Earth can be seen from it at one time or another.

The motion of the launch platform is something else we have to consider, particularly on longer space flights. In our case the launch pad is the Earth, which is not only moving around the sun, but is also spinning on its axis. If you think these motions are negligible, consider that the Earth is moving around the sun with a speed of 66,000 mph and is rotating at more than 1,000 mph (speed of a point on its equator). Space engineers obviously have to take these motions into consideration. If we blast off in the direction the Earth is already heading, we obviously get a big boost, which can save a lot of fuel. This means that if we are trying to launch a rocket to the moon or one of the planets there are certain times that are optimal for blastoff; we refer to them as "windows."

Finally, I should mention that a rocket is coasting most of the time it is in space. Rockets are blasted only briefly—at the beginning of the flight for liftoff and to change orbit once they are in space.

Rockets and Rocketry

Rockets appear to be relatively simple; they consist merely of a long cylinder with fins and a lot of fuel. But appearance is deceptive; they are more complicated than you may think, and they involve a lot of science. Even the rocket fuel is more complicated than you might think. Two types of fuel are used: solid and liquid. The major difficulty with rocket fuel is that it is used in space, and since (like everything else) it needs oxygen to burn, this presents an obvious problem: there's no oxygen in space. The oxygen must therefore be added to the propellant. In solid fuel, oxygen is mixed in with the fuel (the fuel itself is usually a mixture of hydrogen compounds and carbon).

The solid-fuel rocket has a relatively simple structure. It consists of a case, insulation, propellant, a nozzle, and an igniter. In a simple rocket, burning takes place only at the end of the rocket over a small area. But rockets have a greater thrust if the burning takes place over a large area. Because of this, most solid-fuel rocket engines have a hole that runs down the center of the propellant. Burning can then take place all along the hole.

How do we ignite the propellant? It is usually ignited by an electrical current that raises its temperature beyond its combustion temperature. The flames and hot gases generated within the propellant are driven down and out the nozzle at the bottom of the rocket. As can be seen in the diagram (fig. 53), the nozzle has a narrow section that accelerates the gas as it leaves the rocket; this increases the thrust.

The other type of rocket engine uses liquid propellant—either liquid hydrogen or kerosene. The first rocket of this type was built by Robert Goddard in 1922. It has a much more complicated engine, because the propellant and oxidizer have to be kept in separate storage tanks before being combusted. The fuel is combined with the oxygen in the combustion chamber, where mixing occurs when the fuel and oxygen are sprayed into the chamber. The ignition gases escape at the lower end through a nozzle.

Considerable heat is generated by the escaping gases, so the nozzle must be cooled. This is accomplished by circulating the liquid hydrogen (which has a temperature of –253°C) through the region before spraying it into the combustion chamber.

When the rocket takes off, we are faced with another problem. The rocket has to be stable in flight, and without several precautions it won't be. Indeed, it could end up tumbling uncontrollably, which would be a disaster. Two types of systems are used for controlling rockets: active and passive. Active elements are moveable and passive elements are fixed. Also important is the *center of mass* of the rocket. The center of mass is important because all objects, including rockets, tumble around their center of mass. Throw a rock, and as it tumbles in space, it moves about its center of mass. As we saw earlier, the center of mass is the balance point of an object; the center of mass of a ruler, for example, is a point half-way through the ruler at 6 inches. If the object is symmetric it is usually easy to calculate the center of mass, but if it is irregular the center of mass can be difficult to calculate.

In flight a rocket can spin or tumble around one or more of three different axes; we refer to them as the roll, pitch, and yaw axes (fig.

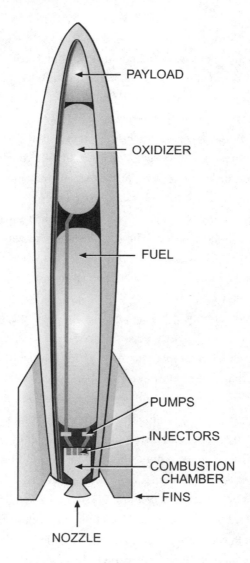

PAYLOAD

OXIDIZER

FUEL

PUMPS

INJECTORS

COMBUSTION
CHAMBER

FINS

NOZZLE

Fig. 53. Schematic of a rocket showing its parts

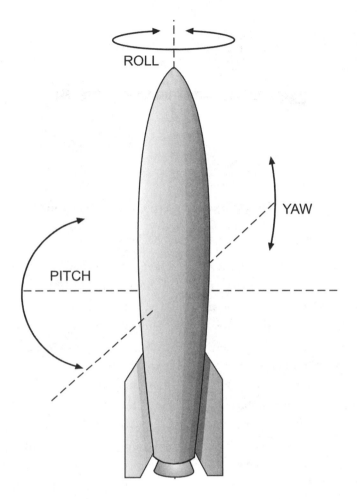

Fig. 54. A rocket showing pitch, roll, and yaw

54). Spin around the roll axis, as we will see, is desirable, but we want to avoid tumbling around the other two axes. Also important for control is what is called the *center of pressure* (fig. 55). It exists only when air is flowing around the rocket (airflow can cause the rocket to begin tumbling), so it's only important during takeoff, not when the rocket is out in space. The center of pressure is defined as the point where the surface area on one side of it is equal to the surface area on the other side of it.

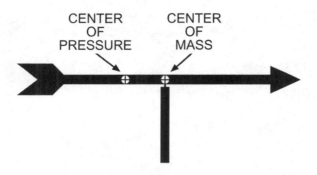

Fig. 55. Position of center of pressure and center of mass

If the rocket has fins at the rear, the center of pressure will be closer to the fins because of their large area. They are therefore important in relation to stability. For maximum stability we want the center of pressure near the bottom of the rocket, and the center of mass near the top. It is particularly important that these two points be well separated; if they are too close to one another the rocket will be unstable.

Not only do we want the rocket to remain stable throughout its flight, we also want to be able to steer it. After all, the rocket wouldn't be much good to us if we couldn't make it go where we wanted it to. This takes us to active or moveable units, and they include moveable fins, vanes, gimbaled nozzles, vernier rockets, fuel ignition, and control rockets. Fins on the inside of the rocket can be tilted to deflect the gas flow; this changes the rocket's direction. Vanes can also be placed in the exhaust of the rocket engines; by tilting the vanes, the direction of the rocket can also be changed. In addition, a gimbaled nozzle (fig. 56) can be used; it can be moved as hot exhaust gases pass through it. When it is tilted in one direction the rocket moves in the opposite direction. Vernier rockets, which are small rockets mounted on the outside, are used to change the orbit in space.

One of the major problems in relation to space travel is mass. There is, in fact, a catch-22 related to mass: the larger the mass of the rocket the greater the amount of fuel that is needed to get it off the

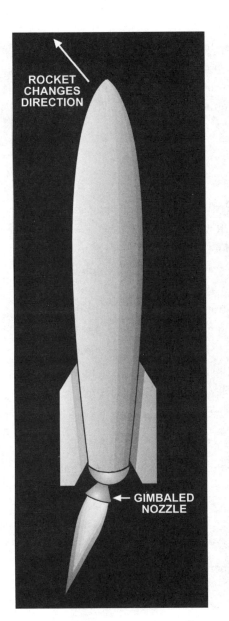
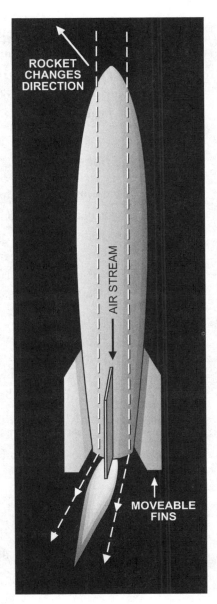

ROCKET
CHANGES
DIRECTION

GIMBALED
NOZZLE

ROCKET
CHANGES
DIRECTION

AIR STREAM

MOVEABLE
FINS

Fig. 56. A rocket showing gimbaled nozzle and fins used to change its direction in flight

ground, but the more fuel it takes, the greater its mass. In practice, about 90% of the rocket is fuel. Of the portion remaining, about 4% accounts for the engine, casing, and so on, and about 6% is payload. In the space shuttle, however, things are not quite this bad; fuel makes up only about 82% of its total weight.

Space Stations

One of the most impressive things in *Moonraker* was Drax's large, spectacular space station. It wasn't anything new, however; space enthusiasts had been talking about the possibility of space stations for years. The first practical design was put forward by Gerard O'Neill of Princeton University in the late 1960s. One of the major difficulties in such a station is lack of gravity, and of course humans need gravity. O'Neill overcame this by designing a rotating station; gravity would be generated at the base of the wheel (via the centrifugal force), and with the proper rotational rate it could easily simulate Earth's gravity. The only place this gravity would exist, however, would be the base of the wheel. Wernher Von Braun had considered a similar idea a few years earlier, but he didn't develop it.

There are other problems in space; one of them is the danger from radiation—particularly cosmic rays. Cosmic rays can be quite deadly, and the inhabitants the space station would need protection from them. Also, the station would have to be big enough to have large windows to let in sunlight; vegetation would not grow without it. And finally both the air and the water would have to be recycled, which might cause a few problems, depending on the size and population of the station.

Moonraker

Moonraker begins with the hijacking of a space shuttle. The shuttle was on loan to the British so, naturally, Bond is sent to investigate. He goes to Los Angeles to visit the manufacturer of the shuttle, Hugo Drax, and he finds him in an elaborate castle (which looks like the Palace of Versailles), and he also soon finds that something isn't right. Drax appears to be up to something (isn't that always the case?).

Most of the early part of the movie takes place on Earth—in exotic places such as Venice and Rio—but I won't say much about this part because we are concerned primarily with space and space travel. Bond soon teams up with Holly Goodhead, one of Drax's astronauts who is presumably on loan from NASA. It doesn't take Bond long to determine that she is actually a CIA agent sent to infiltrate and investigate Drax's organization. They eventually decide to work together.

Bond and Goodhead track Drax to a site in South America. It is from here that Drax plans to launch an armada of space shuttles that are conveniently concealed in the jungle. Both Goodhead and Bond are captured by Drax's men (Goodhead is actually captured first), and they are placed in the area under the shuttle (this is the only part taken from Fleming's book), where they will be "roasted" when the shuttle takes off. They escape, however, through a ventilation shaft (as in the book).

After they escape they manage to overcome two of the pilots of one of the shuttles and take it over. While they are settling in, several other shuttles blast off, including the one with Drax in it, and I must say that the launchings are spectacular and realistic. Goodhead is, of course, an astronaut, so she knows how to pilot the shuttle. (She's really a CIA agent, but I suppose she could have picked up the skills to pilot a shuttle somewhere along the way.)

The fleet of shuttles is headed for Drax's space station, which is one of the highlights of the film. But why wasn't such a large space station seen from Earth? According to the film, Drax had a highly sophisticated radar-jamming system that made the station invisible to radar. In reality, of course, its invisibility to radar would be of little consequence; it was large enough that it could easily have been picked up by telescopes on Earth, and there would be no "jamming" of them.

When Bond and Goodhead docked, they got off and walked around the space station, so it obviously had gravity. But we know there is no gravity in space; it would have to be generated artificially. This has always been a problem for space stations, but as we've seen, it was overcome in the late 1960s by Gerard O'Neill of Princeton University through the use of a spinning space station. Drax's space station is impressive, and you have to give the designers credit, but it doesn't look anything like O'Neill's wheel. It has a large sphere at the center, with spokes out to small docking units. Yet gravity is presumably generated from the rotation of the unit. Strangely, though, the center sphere has floors like an ordinary building and the main floor is not at the outer surface of the sphere, as it should be (this is where the centrifugal force would be acting).

In the film, Bond and Goodhead, now in proper uniforms, attend a speech by Drax in which he outlines his mission. He plans to destroy all human life on Earth with a poison that he has extracted from a rare South American orchid. He will then repopulate Earth with a "perfect race" by using the couples from the shuttles that he has brought from Earth. Drax will rule the Earth. Bond realizes that if he can turn off the radar-jamming system, the space station will be detected on Earth. He makes a break for the control room and manages to turn off the jamming system; as expected, the station is detected from Earth, and a shuttle is sent up to investigate.

Drax discovers what has happened and is ready for the shuttle. He also realizes a saboteur is aboard. Jaws soon captures Bond and Goodhead, but Bond does some fast thinking; he convinces Jaws that Drax is eventually going to do away with him because he doesn't fit in with his "perfect race." Meanwhile, Drax's laser cannons are targeting the approaching space shuttle, and Bond knows he has to do something before it is destroyed. He makes a run for the control room and fires the retrorockets, stopping the spin of the station and destroying gravity on the station. For the next few minutes everyone is floating around in the space station, as they should be.

Drax now sends his men to fight the men coming from the space shuttle, and during the next few minutes an amazing space fight takes place. The men from the shuttle move upward toward the space station, so they have to have jet packs on their backs. The fight is quite spectacular with the flashes of laser guns streaking across the screen, but as I mentioned in an earlier chapter, you can't see the flash from a laser. Of course, the scene would have been much less spectacular if this had been the case.

The fight then moves into the space station, and soon, after all the damage to the station, it begins to break up. Drax heads for one of the exit tubes, but Bond follows him. Just before Bond catches him, Drax grabs a laser gun from the floor and confronts him. "Allow me to put you out of my misery," he says. But Bond shoots him first with one of the darts from his wrist gun; he then pushes him into the airlock and opens the exit door as he says, "Take a few steps for Mankind, Drax." In case you don't recognize this, it's based on the famous words Neil Armstrong spoke when he first stepped on the moon.

By now the giant space station is disintegrating. The producers and so on make the same mistake that almost every other producer has made under similar circumstances. Explosions are going off everywhere, each with considerable noise. But there's no air in space, and therefore no noise.

Earlier Drax had launched three spheres containing a deadly gas toward the Earth. He had planned to launch twenty, and they would have annihilated all human life on Earth. It's obvious that just three could do a lot of damage. Bond and Goodhead run for a shuttle in an effort to destroy the spheres before they get to the Earth. And again this sequence has some scientific problems. First, the spheres had been launched quite a bit earlier, and a shuttle isn't a jet plane. You couldn't simply take after them as you would in a jet plane, unless you wanted to consume an incredible amount of fuel. Space flight is different; you use rockets and retrorockets, usually blasting them only for very short intervals to change orbit. In any case, Bond and Goodhead catch up with the spheres, which are already hitting the top of the Earth's atmosphere. Using the laser in the shuttle, they blast the spheres one at a time. By the time they get to the third sphere it's well into the atmosphere and is starting to disintegrate from the heat. I'd like to know how a laser gun could save the Earth from this poison. All it could do is heat the spheres, and they were already being heated. It is possible, of course, that the laser gun could cause their disintegration, but the gas would still enter the atmosphere. Oh well . . . maybe the heat was high enough to destroy the gas—who knows?

In the final scene Bond and Goodhead are discovered in a "compromising" position. They are presumably naked beneath a blanket. But again, they are in space, and with no gravity the blanket would soon float away. I know . . . it's fantasy; besides, it's a James Bond movie. We should overlook all the scientific inaccuracies, and I can't argue against that. Because *Moonraker* had a space theme, and I'm a great fan of anything involving space and rockets, I thoroughly enjoyed the film. And the special effects were spectacular—no, they were more than spectacular.

Other Space Adventures

Moonraker is not the only film that involves rockets and space. Rockets were in the very first Bond film, *Dr. No.* From his laboratory on Crab Key, Dr. No was interfering with the rocket launches from Cape Canaveral. Rockets were disappearing; they were being brought down by a signal from a device that Dr. No was using.

Rockets also played an integral role in *You Only Live Twice.* The film opens with an American space capsule orbiting the Earth with one of its crew outside the capsule for a space walk. Suddenly a strange spaceship appears and, as it approaches, its nose cone opens like the petals of a flower. It swallows the capsule, severing the astronaut's lifeline, and heads back to Earth.

As Bond later discovers, the mysterious rocket has headed for an extinct volcano in Japan. In one of the scenes a rocket is seen landing in the volcano. In practice this would be quite a feat for a rocket. In another scene a rocketship takes off, and in the final scene the SPECTRE spaceship, with its jaws open, is approaching an American space capsule. But, of course, Bond comes to the rescue and the SPECTRE spaceship explodes after Bond presses the right button on the control panel.

Nuclear Weapons and Reactors

Swoooosh! Bond ducks as Oddjob's lethal hat goes spinning over his head. He is handcuffed to an atomic bomb that will blow in seconds. The key to his handcuffs is only feet away in the pocket of Kisch. Can he grab it before Oddjob reaches him?

The hat cuts a power cable behind him, sending sparks across the room.

Bond grabs for the keys as Oddjob races down the steps. He gets them just in time.

He frees himself, but before he can move Oddjob grabs him and tosses him across the room.

Bond fights back but is no match for the huge Korean. Breaking loose he grabs Oddjob's hat and hurls it at him. Oddjob ducks and it sticks between two metal bars. He moves toward it.

Bond sees his chance. He grabs the exposed power cable and jams it into the bars just as Oddjob touches his hat.

"Aaaah!" screams Oddjob as he falls to the vault floor in a tremendous display of fireworks.

Bond turns to the bomb. It is still ticking away. He has to defuse it. Using a bar of gold he smashes his way into it, but is confused by the large number of wires and gadgets before him.

The bomb continues ticking: 10, 9, 8, . . . Bond grabs some of the wires and pulls, but it continues to click. Only seconds to go!

Suddenly it stops. He looks around. CIA agent Felix Leiter has turned it off.

Wow! That's great stuff. A suspenseful ending to *Goldfinger*. And it's typical of most Bond movies. I wonder what would have happened if the bomb had actually exploded.

I was pretty young when the first atomic bomb exploded so I don't remember my reaction, but I suppose it was surprise and awe. What I do remember, though, is a documentary I saw on the bomb a few years later, and it had a serious effect on me. At that time I hadn't thought of science as a career (I was still trying to decide between being an airline pilot and a car designer), but I know it gave me a lot to think about. Einstein's role in making the bomb, the discovery of nuclear fission and so on were all discussed, but what stood out the most for me was the slow-motion pictures of exploding atomic bombs. I was in awe.

During the next few years most of the interest was in hydrogen bombs. Hydrogen bombs were in the news a lot, as bigger and more devastating ones were built. The United States was in a race with Russia to build bigger and bigger bombs, and soon both sides were stockpiling them. Eventually both sides had enough weapons to blow up the world several times over. I'm not sure why they needed so many, but I guess they were trying to intimidate one another.

Atomic and hydrogen bombs are used a lot in the Bond movies. In most cases they don't say which is being used, so for the most part (unless it's obvious) I'll refer to them as nuclear bombs. Interestingly, the Bond series begins with an atomic bomb, or at least a slowed-down version of it, which is known as a reactor. Dr. No used a nuclear reactor to supply energy to his island.

Nuclear bombs were featured in many of the movies that followed, including *Thunderball, The Spy Who Loved Me, Octopussy, A View to a Kill, GoldenEye,* and *The World Is Not Enough.* So there's no doubt that they play a large role in the movies. Let's look at them in more detail.

The First Nuclear Bomb

How did nuclear bombs come about? Was somebody actually looking for a superbomb, or did someone just stumble on it, as is the case in many discoveries. The key breakthrough came in 1905 when Einstein discovered his famous formula $E = mc^2$. It told us that there was a link between matter and energy; they were, in essence, the same thing, and under the proper conditions one could be transformed into the other. Einstein certainly understood this, but he was equally certain the discovery would not lead to a superbomb—at least not in his lifetime, and he said so many times.

Having an equation that predicts that such a bomb might be possible and actually building one are two different things. Lucky for us, it's not a simple matter to convert matter into energy (if it was, everyone might be trying to build an atomic bomb in their basement). Matter is only converted to energy in certain types of nuclear reactions. To understand how they take place we'll begin with what is called *binding energy.* The nucleus of every atom is built up of neutrons and protons, and when they're very close to one another they're held together by an exceedingly strong force called the *strong nuclear force.* Although it is strong, this force has a very short range; in other words, it only acts over a short distance. Furthermore, it isn't the only force acting in the nucleus; as you no doubt know, the protons have a positive charge, and they therefore repel one another. This force (called the electromagnetic force) isn't nearly as strong as the nuclear force, but it still plays an important role.

The binding energy of a nucleus is the energy (or force) that holds it together. In most nuclei it's so strong, it's impossible to overcome (except in collisions in large accelerators). But as the nucleus gets larger and larger it has more and more protons in it, and protons repel one another. This tends to weaken the overall binding force. The nucleus with the most protons, namely uranium, is therefore of considerable interest, and an experiment that was performed in 1939 showed that it had an amazing property.

The experiment was performed by Otto Hahn and Fritz Strassmann of the Kaiser Wilhelm Institute in Berlin. They bombarded uranium with neutrons. Neutrons are excellent projectiles because they are uncharged, and the nucleus doesn't repel them. The two men were surprised when they discovered barium among the products resulting from the bombardment. This seemed impossible. Barium is only half as heavy as uranium and has only about half as many particles in its nucleus. They performed the experiment again, and again barium appeared in the products. They hadn't made a mistake.

As leader of the group, Hahn wasn't sure what to do. The only person he could turn to for a possible explanation had just left the group. Her name was Lise Meitner, and she had fled Germany a few weeks earlier; she was Jewish and Hitler was beginning to round up the Jews. She was now in Sweden. The only thing Hahn could do was write her, explaining the results, and asking her if she had any idea what was going on.

Meitner was surprised when she received the letter and a little confused. She was also depressed, having barely escaped the Nazis; furthermore, she was now in a foreign land with little opportunity to do research. And, to top it off, it was close to Christmas, and she was lonely. About the time she received the letter, she had gotten in touch with her nephew, Otto Frisch, and was planning to spend Christmas with him. She showed the letter to Frisch, who was also a physicist, the moment he arrived. They talked about it as they went for a walk in the snow. Meitner had a pencil and paper along in case she needed it, and during the next couple of hours she used them.

The only conclusion they could come to was that the uranium nucleus was somehow splitting in half. But it didn't make sense. Was it possible that when it absorbed the neutron it became unstable? They considered what might happen if this was the case. One possibility was that it might begin to oscillate, and if it did, it would eventually begin to look like a dumbbell. If so, the two ends of the dumbbell would repel one another and might eventually break apart. Meitner made some calculations and it was soon clear that this was, indeed, what was happening. Frisch later called the phenomenon *fission*, which means breaking apart.

Frisch passed the information on to the Danish physicist, Niels Bohr, who was just leaving for America. Bohr promised to keep it a secret until they could publish it, but his assistant knew nothing of the promise, and told several people as soon as they docked in New York. The news spread rapidly and rumors were soon rampant, so a formal announcement was made at a Washington conference on physics. The audience was shocked. One of those in the audience was Leo Szilard, an exile from Europe, who had worked with Albert Einstein many years earlier. His first thought was that Germany had the same information and would no doubt attempt to use it to build a superbomb. He talked to Nobel Prize winner Enrico Fermi, who was also at the meeting, and convinced him to talk to the military. Fermi arranged for a meeting with naval officials, but they didn't take him seriously.

Szilard was disappointed, but he wasn't going to sit still; he knew he had to do something. He went to visit Einstein, who was living on Long Island, and they decided that a letter addressed to President Franklin Roosevelt would bring the fastest results. Szilard composed it, and after a few changes, Einstein signed it. It was delivered to Roosevelt by Alexander Sachs, one of the few people who had access to him. To make a long story short, a project called the Manhattan Project was initiated with Enrico Fermi as its director. The first step in building a superbomb was to build a "slowed-down" version of it that could be controlled. This would tell them if it would work. It was built at the University of Chicago, and on December 2, 1942, it was tested—and it was a success. A nuclear bomb was possible!

All they had to do now was build it. A site in New Mexico

called Los Alamos was selected for the design and research. Some of the best and brightest scientists in the land were assembled there. During the next few years all the problems were ironed out and the first bomb was built. It was tested near the city of Alamogordo, New Mexico, on July 16, 1945, and was successful. Within weeks atomic bombs were dropped on Hiroshima and Nagasaki in Japan.

Einstein's equation was the basis of the bomb. It told us that matter could be converted to energy, and the bomb was the proof. But Einstein himself made no other contribution to the project. He was, in fact, a strong pacifist and was against the use of the bomb. When he heard that the United States had dropped one on Hiroshima he was shocked. His only comment was, "Alas, Oh my God!"

Designing a Nuclear Bomb

Let's look at how a nuclear bomb is built. Most of the details are secret, so we won't be able to get down to the nitty-gritty, but the theory behind it is well known. In this section we'll be talking about the atomic bomb (so I won't bother with using the term "nuclear bomb") and, as we will see, the basic idea behind it is relatively simple. It's the engineering details that are complex but we won't get into them. As we have seen, the bomb is based on fission or the breaking apart of an atomic nucleus. Uranium was therefore of immediate interest because it was the largest naturally occurring atom. Uranium, however, has two isotopes (nuclei containing the same number of protons, but slightly different numbers of neutrons; most elements have several isotopes) called U-238 and U-235 (the number is the total number of particles in the nucleus), and only U-235 is fissionable under reasonable conditions. In short, it will fission if a neutron is projected at it, and it absorbs the neutron.

The problem with U-235 is that it is almost exactly the same as U-238 and natural uranium is a mixture of the two isotopes (it is 99.3% U-238 and 0.7% U-235). And because we need pure, or at least highly enriched, U-235, we have to separate the two isotopes. Because of their similarity, they cannot be separated chemically; mechanical methods have to be used. In practice it takes three different methods to get the purity we need. First, natural uranium in gaseous form is passed through miles and miles of pipes that contain holes that are less than a millionth of an inch in diameter. As the uranium passes through them, the U-235 diffuses through the holes faster than the U-238. This helps concentrate the U-235, but it isn't enough. The uranium is then passed through a magnetic devise that enriches it further, and finally it is centrifuged. In the end we have highly enriched (but not necessarily pure) uranium 235.

From here the procedure is relatively simple. We need what is called a *critical mass*. To see why, let's assume the mass we have is in the form of a sphere; it might be the size of, say, a soccer ball. All we need to do to fission a nucleus within this mass is to shoot a neutron at it. Actually, it's even simpler than this: Uranium is radioactive so it gives off its own neutrons. Anyway, let's assume a neutron hits a nucleus and causes it to fission. Energy will be released, but the reaction also produces two or more new neutrons, and they, in turn, will fission other nuclei. These new fissions will produce further neutrons that cause more fissions and so on (fig. 57). This gives what is called a *chain reaction*. We can think of it as proceeding according to the sequence $1 \to 2 \to 4 \to 8 \to 16 \to 32 \to 64$ and so on, and you know how that goes. This is actually a slight oversimplification, but it gives you an idea of what happens. And amazingly, the whole sequence takes only a millionth of a second to complete.

Our aim, then, is to create a chain reaction. But let's look at what happens when the reaction occurs. Will it be self-sustaining? In other words, will it continue until we have a really large explosion? As it turns out, the conditions have to be just right for this. First, if the mass is too small, many of the neutrons will escape through the surface and will not cause fission. If too many escape, the bomb will fizzle. Scientists use an index called k when discussing fission. When $k = 1$ the chain reaction is self-sustaining, but it will not increase in rate. In this case the mass is exactly critical, and it's where we get the term critical mass. If k is

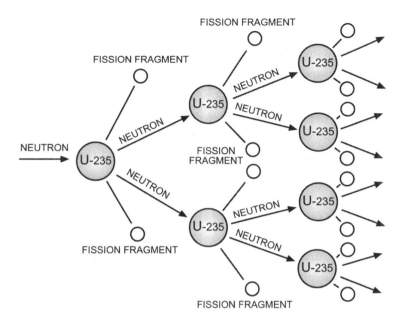

Fig. 57. Schematic of a chain reaction

slightly larger than 1.00, however, the mass is referred to as supercritical and the reaction will increase. This is what we want in a bomb. If k is less than 1.00, on the other hand, it is referred to as subcritical.

On the basis of this, it seems as if we could just keep piling on U-235 and make the bomb bigger and bigger and correspondingly more powerful. But it doesn't work this way. There's another problem. Let's assume the fissioning begins near the center of the mass. Almost immediately the explosive energy released by the fissioning begins to blow the mass apart. Indeed, much of it is blown apart before it is fissioned. This is a serious problem if the mass is very big, and it means that there is also an upper limit to the size of the mass.

In designing a bomb, then, we need a mass that is large enough to keep the fissioning going, but not too large. And you have to be careful; you can't leave a mass of critical size sitting around. It will fission all by itself, and you know what that means. To get around this we have to been a hydrogen bomb. Aside from this, though, hydrogen bombs are a natural extension of our discussion of atomic bombs. Furthermore, literally all bombs now in stockpiles are hydrogen bombs. As we saw, atomic bombs are based on a breaking apart of a nucleus; hydrogen bombs, on the other hand, are based on a coming together or fusing of light nuclei. When light nuclei such as hydrogen come together and fuse, they give off energy. The process is called *fusion*, and it occurs in all stars, including our sun.

The fusion reactions that take place in our sun are exceedingly slow, however, and aren't of much help in building a bomb. We need reactions that are fast—extremely fast. With fusion we are dealing with hydrogen and its two isotopes, deuterium and tritium. In the sun, four hydrogen atoms fuse to form helium, but the process takes a few million years for a given atom, so it's obviously far too slow. We want reactions that take a millionth of a second or less, and, indeed, there are several reactions involving deuterium and tritium that are very fast. The problem is that deuterium and tritium are relatively rare. They have to be separated from ordinary hydrogen. Water can be used as a source of hydrogen, but only about one atom in 5,000 in water is deuterium, and only one in a billion is tritium.

Several fusion reactions involve deuterium and tritium. For example, the direct fusion of deuterium and tritium give helium with a very fast reaction time. But tritium is very expensive to produce; it costs 80 times as much to produce one gram of tritium as it does to produce one gram of Pu-239. An alternative is a reaction between two deuterium atoms. The fusion in this case produces an isotope of helium, and again it is very fast. Other reactions have also been considered, but we won't go into the details.

The preceding reactions do not take place without considerable input of energy or heat, so once we have decided on a reaction we have to consider where we're going to get the needed energy. And here we're lucky: the explosion of a fission bomb produces enough energy to cause the fusion of hydrogen or its isotopes. This means we merely have

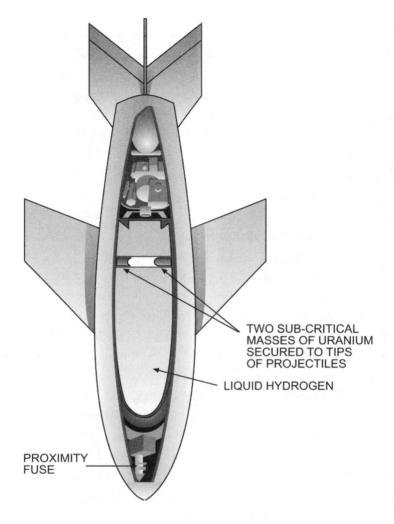

TWO SUB-CRITICAL
MASSES OF URANIUM
SECURED TO TIPS
OF PROJECTILES

LIQUID HYDROGEN

PROXIMITY
FUSE

Fig. 58. Simple schematic of a hydrogen bomb

to surround an atomic bomb with deuterium or tritium, or whatever we are using, for the hydrogen bomb, and it will fuse and release its energy when the atomic bomb explodes. So a hydrogen bomb is not a pure fusion bomb; it also includes an atomic bomb (fig. 58). Actually, in most hydrogen bombs we go a step further. It is well known that U-238 will fission at the high temperatures created by the hydrogen bomb, so in almost all modern nuclear bombs an outer casing of U-238 surrounds the hydrogen.

Is the size for such bombs limited as it is for the atomic bomb? In theory there is no limit; you can make a hydrogen bomb as large as you want, as long as you have the fuel. To give you an idea of the power of nuclear bombs let's look at some of the bombs that have been built and tested. The bomb dropped on Hiroshima was a ten-kiloton bomb (equal to 10,000 tons of TNT); the bomb dropped on Nagasaki was about the same. They caused total destruction for a distance of about a mile around the blast site, and severe destruction out to about three miles. With a one-megaton bomb (one million tons of TNT), severe destruction would extend out to ten miles, and for a 20-megaton bomb, severe destruction would extend out to 35 miles, and we have bombs much larger than this. What is the biggest bomb ever built? The Russians exploded a 50-megaton bomb in the 1950s. The nuclear bombs in the Bond movies were babies compared with the Russian bomb, although in some cases there was no mention of how powerful the bomb was.

You may have heard of cobalt and neutron bombs and wonder why I haven't mentioned them. Well, that's what I'm going to do now. The main difference in these two types of bombs is their goals. The bomb that Goldfinger was going to use on Fort Knox was actually a cobalt bomb. In this type of bomb, cobalt is used in the shell, because natural cobalt is converted to cobalt 60 in the explosion; cobalt 60 is a powerful, long-term emitter of gamma rays. The main purpose of the cobalt bomb is to cause radioactive fallout, which will make the region around the explosion uninhabitable for many years. With the generation of cobalt 60, a region would be uninhabitable for at least five years. Luckily, no such bombs have been built.

Neutron bombs are small thermonuclear bombs that are particularly deadly over a small region. In most hydrogen bombs a shield prevents large numbers of neutrons from being released. In neutron bombs these shields are removed, allowing the neutrons to be released.

Neutrons are much more penetrating than radiation and, in particular, are effective in penetrating shielding that is designed for gamma rays. Neutron bombs do not have a long range, however, so the damage would be limited to a relatively small area.

Effects of a Nuclear Explosion

Although most of the nuclear bombs in the Bond movies are defused in time, a couple of bombs do explode. In *The World Is Not Enough* there is an underground explosion that Bond and Dr. Jones manage to outrun. I have my doubts that anyone could outrun any kind of a nuclear explosion. Fireball velocities can be as high as 700 mph, and then there's the radiation (it travels at the speed of light).

An explosion also occurred in *GoldenEye*, although it was really an EMP (electromagnetic pulse). EMPs do little structural damage; their main purpose is to disrupt electronic and electrical devices. In *GoldenEye*, however, the EMP did appear to do considerable damage. It came from the GoldenEye satellite, but EMPs are also generated in nuclear blasts.

The energy released in a nuclear explosion takes four main forms: blast energy, thermal radiation (or heat), ionizing radiation, and fallout radiation. In general, about 40–60% of the energy comes in the blast, about 30–50% is thermal radiation, about 5% is ionizing radiation, and about 5–10% is fallout radiation. The exact output depends on the type of weapon and its overall energy.

Most of the visual destruction is caused by the initial blast; winds in this blast can be several hundred miles per hour. One of the major effects is the sharp increase in pressure exerted by the shock wave that is created in the blast. The most damage is caused by this wave and the high winds.

Thermal or heat radiation doesn't cause as much destruction, but for people near the blast it can be devastating. Thermal radiation causes severe burns and eye damage; in fact, anyone near the center of the blast will be blinded by it. And even at a distance of several miles, depending on the power of the bomb, many people will be blinded.

The ionizing radiation consists mostly of gamma rays, but high-speed neutrons (which are not radiation) also cause a lot of damage. Neutrons, however, are absorbed relatively quickly, so gamma rays are more dangerous at some distance from the blast. Finally, we have the fallout radiation. It is delayed, but can also be devastating. Much of it is carried by the winds and rain to distant points, sometimes hundreds and even thousands of miles away.

Let's consider the effect of, say, a one-megaton blast, which is small compared with what we have in our arsenal. A one-megaton blast creates a fireball that covers a hundred square miles. (The fireball from a 20-megaton blast, on the other hand, would cover 2,500 square miles.)

In a one-megaton bomb anything within six square miles of the blast will be vaporized instantly. A person in this region will not know what hit him. Anyone within 10 miles of the blast will be blinded immediately, then a pressure wave will hit them. The debris and shrapnel that is carried by the accompanying winds will also strike them at speeds of over a hundred miles per hour. Then, during the next few minutes, they will be hit by a blast of gamma rays, a pressure wave, and finally, a few moments later, they will be hit by winds coming from the opposite direction as the firestorm is sucked back to the center. It's not a place anyone would want to be.

The cloud from a one-megaton blast will reach 10 miles across and 10 miles high, and it will last for about an hour. It is radioactive and will send radioactive material into the upper atmosphere where it will spread for hundreds or possibly thousands of miles downwind.

Even if you survived the initial blast, your problems would not be over. Almost everyone will have received a high dose of radiation and will later develop cancer, leukemia, or other diseases.

All this destruction from one bomb, and we still have a stockpile of more than 10,000 nuclear weapons. At one time we actually had more than 40,000, and the Russians likely had an equal number. Our real danger now, however, is that many other nations have built, or at least are in the process of building and stockpiling nuclear weapons. China, France, Israel, the United Kingdom, India, and Pakistan all now have stockpiles. What is really worrisome is that North Korea now also appears to have several bombs. Besides this there are numerous wannabes (nations hankering for nuclear weapons). They include Egypt, Syria, Iran, Taiwan, South Korea, and a few others—and they may, indeed, eventually get them. We should encourage them to save their money and just watch Bond films.

Nuclear Reactors

As we saw earlier a nuclear reactor is a slowed-down nuclear bomb. In this case the energy doesn't all come at once, so it is usable. Hundreds of nuclear reactors across the United States and around the world supply energy for electric lights and other power needs. (So nuclear energy does some good, after all.)

The reactor operates as a result of the fissioning of U-235 or Pu-239, but in a reactor, as compared with a bomb, we need complete control of the process. And for this, the first thing we need is a *moderator;* a moderator is a material that slows down the neutrons that are emitted (fig. 59). The best moderators are graphite and heavy water. Slowed-down neutrons work better because they are more likely to fission a uranium or plutonium nucleus when they strike it. But we need more; we have to keep the reactor under control once it gets going. We can't let it get out of control, or we'll have a bomb. This is done with control rods; they are made of material that absorbs neutrons, with the two best materials being cadmium and boron.

In a reactor we have uranium or plutonium in the form of rods, surrounded by water or other moderator. The control rods are interspersed throughout the reactor. A reaction is initiated by pulling out one or more of the control rods. By adjusting them the reaction can be speeded up or slowed down. The water in the reactor acts both as a

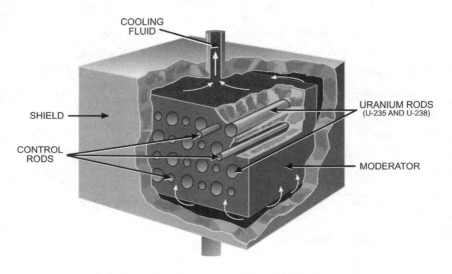

Fig. 59. Schematic of a nuclear reactor

moderator and a coolant that controls the temperature of the core and prevents the fuel from melting. The system is usually under pressure, so the water is superheated. The superheated water drives a steam generator that, in turn, drives a turbine that creates electricity.

Several different types of nuclear reactors actually exist. The one discussed here is referred to as a boiling-water reactor. Another one of considerable importance is the breeder reactor. It is a reactor that produces Pu-239 from U-238 and therefore generates its own fuel.

Nuclear Weapons in the Movies

Nuclear weapons play a large role in the Bond movies. Indeed, the very first one, *Dr. No*, has a nuclear reactor in it. We see it toward the end of the movie. After Bond and Honey Ryder are captured, Bond escapes and manages to make his way to Dr. No's lab, where the reactor is at the center of the room. It uses water as a moderator.

Bond overpowers one of the technicians and puts on his suit, then he goes to the catwalk above the reactor pool. The countdown has begun for an attack on a rocket that is being launched from Cape Canaveral; the rocket can be seen on several of the TV screens in the lab. Bond waits until the last moment, then swings into action: he turns a wheel on the reactor control panel that speeds up the reaction and causes it to go out of control. (I'm not sure how he knows this wheel controls the reaction level, but I guess that's not important.)

A technician tries to stop him, but Bond quickly overpowers him. The reactor is now out of control and an alarm sounds. Everyone in the room rushes for safety, but Dr. No heads for Bond, who is still on the catwalk. He strikes out with his mechanical hands at Bond, but Bond manages to elude them. Suddenly both Bond and Dr. No fall to a platform beneath the catwalk that is slowly sinking into the furiously boiling water below. Bond manages to escape, but kicks Dr. No back onto the platform, where he slowly disappears into the boiling water. In the final scenes, Bond and Honey escape and the island blows up.

One of the difficulties with the reactor scene is that the reactor gets hotter when the control rods are inserted into the core. It should get hotter when they are withdrawn.

We have already talked about the bomb in *Goldfinger*. We saw that it was a salted or cobalt bomb that was designed to produce a large amount of radiation. Goldfinger wanted to make the gold at Fort Knox radioactive, and therefore unusable for years, so that his gold would increase in value. A strange plot, but it makes more sense than trying to steal the gold, which would have weighed hundreds of tons and taken weeks to move.

Atomic bombs played an important role in *Thunderball*, but we don't see much of them. SPECTRE agent Largo hijacks a NATO bomber with two nuclear bombs aboard. He then asks for a ransom of 100 million pounds of uncut diamonds. If the ransom is not met he will unleash the bombs on two unspecified cities. Bond is brought in to locate the bombs and stop SPECTRE from using them.

The bomber is brought down in the ocean near the Bahamas, and covered with netting so that it can't be seen from the air. The bombs are taken to a cave. Toward the end of the movie, Largo and his frogmen go to the cave and load the bombs on an underwater saucer and take them to Largo's yacht.

Bond is working with CIA agent Felix Leiter, and Leiter is waiting with a team of frogmen. Bond finally finds out the target is Miami and radios Leiter. As Largo heads toward Miami the Navy frogmen intercept them, and a spectacular underwater battle ensues. Largo escapes in his yacht with one of the bombs, but Bond manages to cling to it and climbs aboard. In the end Largo is killed and the yacht crashes, with Bond and Domino escaping just in time.

Both nuclear bombs and nuclear reactors play a role in *The Spy Who Loved Me*. Two nuclear submarines are captured and stored inside Stromberg's huge supertanker *Liparus*. Nuclear submarines are, of course powered by nuclear reactors and they are equipped with missiles that have nuclear warheads. Toward the end of the movie the two nuclear submarines, manned now by Stromberg's men, leave the tanker to go to predetermined positions where one will fire a nuclear missile at the United States and the other will fire one at the USSR. This will presumably trigger World War III.

Bond was on one of the captured submarines, so he is now aboard *Liparus*. Before the two submarines can fire their missiles he breaks into the control room and reprograms them. When they are fired, therefore, they don't head for the United States and the USSR, but they pass one another in flight and come down on one another (an idea taken from Fleming's book *Moonraker*).

A nuclear bomb is also used in *Octopussy*. It was taken by train to a U.S. airbase at Feldstadt, West Germany, where Octopussy's circus was scheduled to perform. Like the other bombs in the Bond movies it looked like a large suitcase, but presumably had enough power to devastate an area within a radius of 20 miles. Bond is on the train hiding in a monkey suit. He watches Gobinda set the bomb's detonator to go off in four hours. Before he can do anything with the detonator, he falls from the train in a fight and has to hitchhike a ride to Feldstadt. He is chased through the base but eventually disguises himself in a clown outfit. Then, with only seconds to spare, he disarms the bomb.

A large bomb, which was presumably a nuclear bomb, was also used in A *View to a Kill*. The villain, Zorin, wants to corner the world's microchip market. He plans on doing it by exploding a bomb under Silicon Valley, which is the source of most of the world's microchips. The bomb is supposed to trigger a massive earthquake, since it is going to be detonated near the San Andreas fault.

In *GoldenEye* General Ourumov and Xenia Onatopp fly a helicopter to the Severnaya weapons facility in Siberia. Ourumov gets the GoldenEye, and to her delight Xenia gets to machine gun everybody. The general then activates the GoldenEye satellite to emit an EMP. Having armed the satellite, Xenia and the general escape in the helicopter. Unknown to them a girl, Natalya Simonova, has escaped by hiding in a cupboard. After they leave she creeps out of the cupboard and surveys the dead. As she looks around, however, the satellite directs an EMP at the base. It hits with considerable devastation. EMPs, however, are not supposed to create a lot of damage; they only damage electronics. Actually, a MIG fighter that is sent out to investigate is also hit by the EMP, and crashes into the site, so some of the damage is done by the crash. In the end the base is completely demolished.

Finally, we have the underground blast in *The World Is Not Enough*. It is presumably an atomic blast because Russian nuclear physicist Christmas Jones tells Bond that there is only plutonium below ground. She also mentions that there is tritium above ground, presumably from a hydrogen bomb explosion. Bond takes the elevator down to the test area, and who does he meet, but his old enemy—Renard. Bond soon has a gun on him, but the tables are quickly turned and Bond becomes the captive.

Meanwhile Dr. Jones arrives on the scene. Within minutes Renard and his men shoot everyone, and Bond and Jones run for their lives. Renard tries to seal them in the tunnel, but he doesn't succeed as Bond dives through the closing door. After a gun battle, Renard escapes and arms the underground bomb.

The bomb goes off and Bond and Jones must run for their lives. They manage to close one of the tunnel doors behind them as the fireball races along the tunnel. Then they take an elevator to the surface and the fireball follows them, and again, they barely escape. A lot of narrow escapes, but that's to be expected. After all, how exciting would a Bond film be without several narrow escapes?

Water Sports and Guns

"What do you think of the Beretta?" M asked Q.

"A ladies gun, sir," said Q.

Bond frowned.

"Why do you say that?" asked M.

"No stopping power . . . appeals to the ladies," said Q.

"I don't agree," said Bond. "I've used it for fifteen years . . . and until now I've had no problems."

This scene is from Fleming's novel *Dr. No.* After Bond's Beretta misfired, M insisted that he change to a new gun, and Q suggested the Walther PPK. Bond was reluctant at first, but he soon realized that M was right, and he ended up using the PPK throughout the next seventeen movies. (A similar scene happened in the movie version but Q was not present.)

The 9-mm Walther PPK was first used by German plain-clothed policemen in the early 1930s. It was flat, small, and easy to conceal, and it had a six-round magazine. By the 1990s, however, it was becoming outdated and when Bond lost it in *Tomorrow Never Dies*, Wai Lin supplied him with a Walther P99. It was a more modern gun and

Fig. 60. Bond's gun, the Walther PPK

---

had several features that the PPK did not have. Like the PPK, however, it used 9-mm ammunition, but it had a magazine that held 16 shells rather than 6. This made it slightly heavier than the PPK when fully loaded (fig. 60).

Bond used many other guns, but the PPK and P99 were his mainstays. Some of the other guns were rifles. In *From Russia with Love*, for example, he had an AR-7 rifle in his briefcase with the barrel, grip, magazine, and telescopic sight fitted into its stock. Agent Kerim Bey borrowed it to kill the Bulgarian hit man Krilencu; he shot Krilencu as he was trying to flee his hideout through a secret door. Bond later used the same rifle to shoot down a helicopter.

In *The Living Daylights* Bond saw the heroine, Kara Milovy, pointing a rifle at General Georgi Koskov and shot it out of her hands. It scared the "living daylights" out of her (and that's of course where the name of the movie comes from). Agent Saunders reamed Bond out for not killing her. "I only shoot professionals," said Bond. "That girl didn't know one end of the rifle from the other." And he was proven right when it was later shown that the rifle was loaded with blanks.

Bond also used an interesting-looking gun in his attempt to shoot Franz Sanchez in *Licence to Kill*. One of the most amazing of the Bond guns, however, appeared in *GoldenEye*. Bond used it to shoot a piton with an attached cable into the concrete of a Siberian dam. He winched himself down to the dam using the cable, then used a laser in the gun to get through a steel door.

Everyone, including Bond, used laser guns in *Moonraker*. The blue flashes looked impressive but weren't very realistic. A "ski pole" gun was used in *The Spy Who Loved Me*. Bond used it to kill Anya's Russian boyfriend (she later vowed to kill him because of this). And finally, in several of the latter movies Bond used submachine guns of various types, but we won't say anything about them.

Physics of Guns

Bullets are, of course, basic to guns, so let's begin with them. I'm sure most of you have a fairly good idea of how they work, but maybe I can clear up some misunderstandings. The basic component of a bullet is, of course, gunpowder, which explodes when ignited. The explosive reaction starts at one end of the gunpowder and travels through it in a fraction of a second. The result is the production of gases that take up a much greater volume than the corresponding solid materials. The sudden increase in volume produces a tremendous pressure, and assuming the reaction takes place in a cartridge, this pressure drives the projectile out of the barrel. The bullet accelerates down the barrel of the gun, and, in general, the longer the barrel (up to a point) the greater the muzzle velocity. The speed with which it exits depends, therefore, on the amount of explosive material in the bullet, the mass of the projectile, and the length of the barrel. Muzzle velocities range from about 3,200 ft/sec in high-powered rifles down to less than 1,000 ft/sec. You can increase the muzzle velocity by adding more gunpowder, but there is a limit; modern guns are designed to withstand pressures of only about 70,000 psi (pounds per square inch). I should also mention that a rifle barrel is grooved so that the bullet spirals down it and emerges with a spin. This gives it more stability.

One of the major problems in firing a bullet is initiating the explosion safely. The technique that is now used was devised by Alfred Nobel in the 1860s. His difficulty wasn't with bullets, but it was basically the same problem. The standard explosive of the time, namely nitroglycerine, was highly sensitive and exploded easily; it was therefore difficult to handle. Nobel found that if he mixed nitroglycerine with diatomaceous earth (silicon skeletons of microscopic diatoms) it was completely safe to handle. He called the new material dynamite. He then found that he could set off the dynamite with a "detonator"—a small amount of a different explosive (called the primer). This is the technique used with bullets. In this case a small amount of primer is placed at the base of the bullet that is set off using a hammer or firing pin, and it causes the gunpowder to explode.

Incidentally, when I refer to gunpowder, I'm not necessarily referring to black gunpowder, which consists of a mixture of sulfur, saltpeter, and charcoal. It was the original gunpowder, but most gunpowder today is based on nitrocellulose or other smokeless explosives.

Our discussion so far has been about the flight of the bullet within the gun, and it is usually referred to as *internal ballistics*. Once the bullet leaves the gun, however, we are dealing with what is called *external ballistics*, and in the next section I'll go into it in considerable detail. To be complete, I should mention that when the bullet strikes the target we are dealing with what is called *terminal ballistics*, but that is outside the scope of this book.

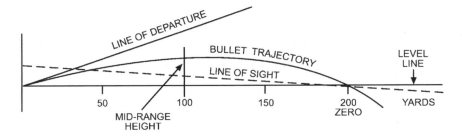

Fig. 61. The bullet trajectory from a rifle

External Ballistics

External ballistics deals primarily with how much a bullet drops once it leaves the barrel of the gun. If you aim directly at something, with the sights aligned along the barrel (the line of bore of the gun), as you likely know, the bullet will hit below the target. The reason, of course, is that gravity is acting on the bullet. Gravity is something we have no control over; it acts on everything regardless of its speed, shape, or the medium through which it is passing. In a given time, any object falls a certain distance (neglecting air friction), and that's all there is to it.

Let's assume we are firing a rifle horizontally. A question that immediately comes to mind is: how long will it take for the bullet to hit the ground? A simple experiment gives the answer, and the result may surprise you. At the same time you fire the gun, drop a bullet to the ground; both bullets will strike the earth at exactly the same moment.

You may think that if you point the gun at an upward angle and fire it, that the bullet won't drop as fast. But it does; it drops at exactly the same rate. Overall, if you neglect air drag on the bullet, it will trace out what is called a parabolic curve (fig. 61).

The basic formula governing the drop is

$$d = \tfrac{1}{2}gt^2$$

where d is the distance of the drop, t is the time, and g is the gravitational constant. But we would also like to know how far along the horizontal the bullet travels in time t. It is given by

$$d = vt.$$

As a simple example, assume we have a bullet that has a velocity of 1,000 ft/sec, and we want to know how far will it drop in 0.25 seconds? Substituting in we get

$$d = \tfrac{1}{2}(32)(0.25)^2 = 1 \text{ ft.}$$

Using the other formula we find that the bullet will travel 250 feet along the horizontal in this time ($1,000 \times 0.25$).

Because of this drop, the sights on a rifle have to be adjusted. They can't be properly sighted in for all distances, however. With a given setup they are accurate only for one particular distance (more exactly, there are two) (fig. 61). We sight in by aligning the bore of the gun above the target. The sights are still pointed at the target, but the gun shoots slightly high, and the bullet traces out a looping curve, coming down across the point where your gun is pointed at a particular distance. This distance may, for example, be 100 yards. For hunters, of course, this is a problem. The game they see is not always at 100 yards; furthermore, it may be difficult to determine exactly how far it is away. Rangefinders (devices for determining the distance to the game) give them a way around this. They tell the hunter how far away the game is, and he can adjust his sights accordingly.

Everything would be great if this was all there was to it. But we have another problem, and it's quite serious. Air resistance causes a significant drag on the bullet. Because of air resistance, the velocity of the bullet slows down during its flight; several things affect this slowdown. In addition, we have to take into account the *inertia* of the bullet. Inertia is the resistance to change in motion, and it depends on the mass (weight) of the bullet. Heavier bullets have more inertia and will be less affected by drag.

Cartridge	Muzzle Velocity (ft/sec)	Velocity (100 yds)	Velocity (500 yds)
.22	2,690	2,042	841
.30-30 Winchester	2,390	1,973	973
.308 Winchester	2,750	2,743	1,664
.30-06 Springfield	2,600	2,398	1,685

So how do we take these things into consideration? We use what is called the ballistic coefficient (BC). Ballistics engineers have experimented with bullets of many different shapes and weights and have come up with what is called the *standard bullet*. All other bullets are compared with it. If the bullet slows in exactly the same way as the standard bullet, it has a BC of 1.0. As you might expect, most bullets have BCs that are considerably less than this; a typical value is 0.300 to 0.400. Anything above 0.400 usually has a streamlined nose profile; roundnose bullets, on the other hand, typically have BCs less than .300. In general we can say that the higher the BC, the better the aerodynamics of the bullet and the less the drag. You have to be careful, though, because BCs change slightly with velocity and altitude, and furthermore, the way BC is determined is not an exact science.

Another term, called sectional density (SD) is also of importance in relation to bullets. It is related to BC through the formula BC= SD/I where SD is M/d^2, d is the diameter of the bullet and I is a factor that depends on the shape of the bullet (it decreases with increased pointedness of the bullet). Bullet makers usually list both SD and BC for their bullets.

The drag (D) on a bullet depends on many factors, including the shape of the bullet, the density of the medium through which it is passing, the diameter of the bullet, its velocity, and a factor that depends on the "rotational" forces on the bullet while it is in flight (e.g., yaw). Drag is related to retardation through the formula

$$r = D/M,$$

where r is retardation.

Bulletproofing

Several of Bond's cars were bulletproof, and Brad Whitaker, the arms dealer in *The Living Daylights*, wore a bulletproof mask and vest in his shootout with Bond. A bulletproof vest is also seen in Q's lab in *Goldfinger*. This brings us to the question: What types of materials stop bullets? Armor plating, which usually consists of reinforced sheets of steel, was used in the cars. Bulletproof glass was used for the windshield of the car and for Whitaker's mask. Bulletproof glass is made by layering a polycarbonate material between layers of ordinary glass. Polycarbonate is a tough, transparent plastic known commercially as Lexan, Tuffak, or Cyrolon.

Whitaker also used a bulletproof vest. Bulletproof vests are now quite common, although to my knowledge Bond never used one. The first bulletproof vests became available in the 1940s and were made of nylon, supplemented with plates of fiberglass, ceramic, or titanium. They were uncomfortable and unreliable. In 1966, however, a chemist at Du Pont invented Kevlar, a liquid polymer that could be spun into fibers and made into cloth. It was originally used for tires and ropes and in boats and planes, but in 1971 someone realized it was so strong it could be used in bulletproof vests, and it has proved to be one of the best materials we have for this purpose. It does have a competitor, however; in 1989 Allied Signal developed what is called Spectra, and it is now used extensively in bulletproof vests, as well.

Today, bulletproof vests are made up of many layers of Kevlar (or Spectra), and it is this layering, or lamination, that makes the vest bulletproof. When a bullet hits the vest it is caught up in a network of very strong fibers that absorb and disperse the energy of the bullet, causing it to flatten or deform. Energy is absorbed by each successive layer of the material until the bullet is finally stopped. Note, however, that no bulletproof vest is 100% effective against all bullets.

It would be interesting to consider how much force someone feels when wearing a bulletproof vest that is struck by a bullet. In the movies we frequently see people being knocked over when they are struck by a bullet. Does this really happen? It's easy to show that the force on a person's body is given by the momentum of the bullet (mv) divided by the time (t) the bullet takes to come to a full stop after hitting the target.

$$F = mv/t$$

The momentum of the bullet is relatively small because it has a small mass. The decisive factor, therefore, is t. t is relatively small for a person without a bulletproof vest, but the bullet usually penetrates far into the victim's body, so it is not nearly as small as when it is stopped by a vest. As it turns out, the force is usually relatively small for a person not using a vest (so it's not likely he'll be knocked over), but this force can be significant for a person with a vest (where the stopping distance, and time, may be a hundred times less).

Water Sports

Boats and submarines are used in several of the Bond movies. Let's look at boats first. And I'm sure you'll agree with me that one of the most impressive of them was the Q boat (fig. 62), named for Q, in *The World Is Not Enough*. It was Q's pride and joy. (He said he planned on using it for fishing when he retired. What was he planning on do-ing—machine gunning the fish?)

Q was showing the boat to Bond when a bomb blew a hole in the wall of MI6 headquarters. Through the hole Bond saw a woman standing at a machine gun in a white boat on the river below. It was the same woman who had offered him a cigar earlier. Bond raced for the Q boat, and after a spectacular jump from the building he took after the white boat with Q yelling after him, "Stop, Stop . . . it isn't ready yet!"

The chase was on, with the much smaller Q boat racing after the white boat. The Q boat was well armed, but surprisingly, so was the

Fig. 62. The Q boat

white boat, and the "cigar girl" tried to machine-gun Bond as he approached. It didn't help, though; he hit her boat and sailed right over it, but it didn't do much damage and the chase continued.

The white boat raced toward a bridge that was closing and just managed to make it. It was closed by the time Bond got to it, but he had a secret weapon: the Q boat submerged and went under it. The cigar girl then headed for an area that contained barrels of fuel; hitting them with considerable force, she caused a huge explosion, and in the confusion she got away. But Bond soon picked her up on his navigator and took a short cut through the streets; at one point, he drove through a café. Jumping back into the river he caught up with her and sent two torpedoes after her boat.

Seeing the torpedoes, she abandoned her boat and jumped into a hot-air balloon. Bond wasn't finished, though; he grabbed onto a dangling rope as the balloon began to rise. Holding onto the rope he yelled, "Who is behind this? Tell me and I won't shoot." But she would not surrender; she shot at the fuel tanks below her and blew herself up.

Bond let go of the rope and fell onto the roof of a large building and rolled down it to safety.

An exciting opening for any film, and typical of most Bond films. Let's look at the Q boat in a little more detail. It had a maximum speed of 100 mph, an acceleration of from 0 to 60 in 6 seconds, a 5.7-liter engine with 300 hp, and an auxiliary jet engine that could be used for extra thrust. At 15 feet 6 inches, it was small, but it was exceptionally well equipped. It had two torpedoes, which could be targeted by using a scanning range finder in the bow. It had a machine gun and rocket launchers, a GPS satellite and radar-tracking systems, and grenade launchers; it was powered by a turbine jet drive. It could operate in just three inches of water and even worked well on dry land (for short distances). All in all, it was quite a boat.

Another spectacular boat chase took place in *Live and Let Die*. Jet-powered boats were used in this chase through Florida's bayous. In fact, Bond's and his pursuers' boats all had Evinrude jet-propulsion motors. The chase in this case, however, was a little overdone with boats hurtling over roadways, skimming across land and swimming pools, and even through an outdoor wedding ceremony. Comedy was added by the local sheriff, J. W Pepper, who got involved in the chase.

And finally there's one of my favorite boat-chase scenes, which happens in *From Russia with Love*. Bond and Tatiana escape in a speedboat into the Gulf of Venice, and it seems that they are finally free at last. But SPECTRE is waiting for them with a fleet of black boats (the villains always use black cars and boats), each armed with machine guns and grenade launchers. The chase is exciting, but when the spare fuel tanks in Bond's boat are punctured by machine gun fire he lets them fall into the water and surrenders. It looks like the end for him. But as usual, he has something up his sleeve. As the black boats approach and prepare to board his boat, he grabs a flare gun and fires it at the fuel floating on the water. Several explosions follow, and in the confusion he and Tatiana escape.

Bond also uses a speedboat in *Moonraker* to get to Drax's headquarters in the jungle. Again his boat had the usual arsenal of gadgets. And again, he is soon being chased by a flotilla of speedboats equipped with mortars. After blowing up two of the boats, he comes to a waterfall; there seems to be no escape. But, surprisingly enough, he pulls out a hang glider and takes off. (I wonder how he knew he would need a hang glider.)

Physics of Speedboats

Most of the speedboats Bond used were jet boats, but the one he used in *From Russia with Love* was propeller driven. So let's start with that boat. How, exactly, does a propeller push the boat forward? It's easy to see from its shape that as it rotates it pushes water backward, or astern. At the same time, however, water rushes in to fill the space on the opposite side of the propeller blade, and this causes a pressure difference on the two sides: a positive or pushing pressure on one side and a negative or pulling pressure on the other. This occurs in all the blades, for the full circle of rotation. The propeller is therefore both pushing and pulling the boat through the water. The overall result is that the boat is accelerated by the motion of the propeller, and in the process, it creates a jet of high-speed water behind the propeller. Physics comes into the picture through Newton's third law; as the propeller pushes water back, it causes a reactive force via Newton's third law that thrusts the boat forward.

In any unyielding medium, each revolution of the propeller would thrust the boat forward by a distance equal to the pitch (angle) of the propeller. But water is not unyielding; it gives under the pressure of the propeller blades and because of this, it pushes the boat forward only about 60 to 70% of the pitch on each revolution. The difference from 100% is called the *slip* of the propeller.

Propellers are interesting, but most of the boats used in the Bond movies are actually jet boats that have no propeller. Let's turn to them. We talked about jet engines in airplanes earlier, and the principle is the same. A jet boat is fitted with an engine that drives a water pump (sometimes called a "jet pump"), which brings water in through an intake in the bottom of the hull, then forces it out as a high-pressure jet at the rear of the boat. The boat is propelled and steered by the thrust of this jet stream.

Again we have the action-reaction principle. The action is the force of the jet out the rear of the boat. The boat is propelled forward by the reaction force according to Newton's third law. You feel the same reaction force on your hands when you hold a garden hose with water "jetting" out of it.

Underwater Devices in *Thunderball*

When it comes to underwater devices we find the largest number in *Thunderball*. But I guess that is to be expected because most of the latter part of the film takes place under water. *Thunderball* was, in fact, the first movie to feature an extended underwater battle. The battle was underway when Bond arrived with his motorized backpack. The backpack allowed him to swim through the group at high speed, ripping off masks and causing lots of chaos. In one scene he tricks two of Largo's men into following him into a wreck on the seafloor. Then, when they are inside, he quickly exits and throws a grenade through the hatch.

In the finale Bond chases Largo and his atomic saucer (which contains the atomic bomb) to his yacht. Largo takes off, but Bond clings to the yacht and manages to climb aboard where he faces and overcomes Largo—with a little help from Domino (she shoots him in the back with a spear gun).

Numerous interesting devices are used in the battle scene, including motorized underwater sleds and the much larger atomic bomb sled designed by Largo. It was housed in the belly of his yacht and was equipped with headlights and armed with six forward-firing spear guns. Indeed, the spear guns used by Largo's men were different in that they had compressed-air canisters attached to them that extended their range to about 24 feet.

Bond's underwater propulsion unit was also well equipped. It was armed with spear guns and a searchlight, and Bond had his handy miniature air tank that gave him 4 minutes of breathing time. He used it in the latter part of the battle when he lost his tank, but somehow it seems to have lasted a lot longer than 4 minutes.

Largo's luxurious 100-ton yacht, the *Disco Volante*, was also spectacular. It contained a special room in its hull that allowed Largo and his crew to hoist two atomic bombs into it and store them. It had many extras, including TV cameras on the underside to watch for intruders, and they caught Bond when he tried to examine the hull.

Submarines

And finally, we can't forget submarines. A couple of interesting ones were used. A minisubmarine called the *Neptune* was used in *For Your Eyes Only*. Timothy Havelock used it in his search for the *St. Georges*, and his daughter Melina, along with Bond, used it to go to the wreck and bring back the ATAC transmitter (control unit for British Polaris subs). It was almost sunk when a one-man sub attacked them. Bond and Melina were captured in the process.

The *Neptune* had a length of 23 feet and was 8 feet wide; it had a two-person cabin with a large viewing window and headlights to increase visibility in murky water.

A much smaller sub called the Bath-o-sub was seen in *Diamonds Are Forever*. It belonged to Blofeld and he tried to escape from the oil rig in it, but it never reached the water. It was a small, one-man sub with a fiberglass hull and searchlights.

A submarine car, the Lotus Esprit, was used in *The Spy Who Loved Me*, but we already talked about it, so I won't say any more about it here. And I think that covers most of what I wanted to say about boats.

Appendix

Bond fans have their own ideas about which Bond film is the best, and how each one rates in general. Let me say immediately that I enjoyed all the Bond films and coming up with the "best" in various categories was difficult. Here, I've put together lists giving you my favorites in nine different areas. See whether you agree.

Best Films (in descending order)

1. *Goldfinger*
2. *From Russia with Love*
3. *The Spy Who Loved Me*
4. *For Your Eyes Only*
5. *The Living Daylights*
6. *The World Is Not Enough*
7. *On Her Majesty's Secret Service*
8. *Thunderball*
9. *Tomorrow Never Dies*
10. *Dr. No*
11. *Licence to Kill*
12. *GoldenEye*
13. *Die Another Day*
14. *Octopussy*
15. *Moonraker*
16. *Diamonds Are Forever*
17. *You Only Live Twice*

18. *Live and Let Die*
19. *The Man with the Golden Gun*
20. *A View to a Kill*

I didn't have much trouble with my first choice. I've watched *Goldfinger* several times and have read the book at least twice, and I'm convinced it is the perfect Bond film. Sean Connery was excellent as Bond, and the rest of the cast played their roles well. Goldfinger was a realistic villain, and his henchman Oddjob seemed invincible. And the first of a long list of Bond cars was used—it's still one of the most talked about cars in history.

From Russia with Love is perhaps a little more realistic and down to earth than *Goldfinger*, but it lacks some of the excitement. Bond relies on his skills and cunning to outwit the villains. He only has one gadget, the briefcase, but he makes good use of it. The plot is brilliant but not overdone, and Red Grant has to be one of the top villains in the series. The fight scene between Grant and Bond on the Orient Express is one of the best in the series; it has a lot of suspense in it. Also, one of the best boat-chase scenes is in this movie.

Roger Moore's first two Bond films (*Live and Let Die* and *The Man with the Golden Gun*) were a bit of a letdown after the earlier films. I liked Moore but was unsure he was the ideal Bond. Then came *The Spy Who Loved Me*, which I now consider to be one of the best films in the series. Moore finally appeared comfortable with his role, and he was believable as a secret agent. The film gave new life to the series. Overall it was spectacular and a delight to watch.

We now come to one of my favorites, *For Your Eyes Only*, which again has Roger Moore as Bond. I would like to have placed it higher but knew that, as good as it was, it was no match for the first three. It has everything: beautiful scenery, excellent skiing and underwater scenes, and a tremendous finale. I particularly like the Citroën chase but have to admit that it wasn't as suspenseful as it was amusing. All in all, though, this film was a good change after *Moonraker*, which had far too much slapstick in it for my taste.

In my next choice, *The Living Daylights*, we have a new Bond, Timothy Dalton. With him came an abrupt change in the way Bond was portrayed. Dalton was much more serious and much less humorous than Moore or Connery. He has been criticized for this, but I'm convinced that the criticism is unwarranted. To me he was a very believable Bond. When he was in trouble, you knew he was in trouble, and worried about how he would get out of it. As far as I'm concerned, *The Living Daylights* is one of the best adventure films of the series. It keeps you on the edge of your seat throughout.

I haven't selected any of Pierce Brosnan's films in the top five, yet Brosnan is certainly one of my favorite Bonds. With Brosnan came more action, his best is *The World Is Not Enough*. Brosnan is, without a doubt, one of the most suave, handsome, tough, and yet funny of the actors that have played Bond. The ski scenes with the parahawks were superb, and the Q boat chase was one of the best boat scenes in the series.

Each of the Bond actors had his own personality, and all were good actors—except one. George Lazenby had almost no acting experience when he was asked to play Bond in *On Her Majesty's Secret Service*. And it showed. You have to give him credit, though, for excellent action scenes, and he didn't use a double in many of them. One of the reasons I've placed *On Her Majesty's Secret Service* so high on the list is the excellent ski scenes. But this certainly isn't the only reason; the plot, story line, and scenery are also excellent. It is also one of the few Bond films that follows Ian Fleming's novel almost exactly, and it was one of his best books. If Connery had been in the lead, it could have been one of the top Bond movies.

Speaking of Connery, we come to another film I would have liked to place higher: *Thunderball*. It had everything you would expect in a Bond film: a cold-hearted villain, a beautiful heroine, beautiful scenery, and lots of action. It was the first Bond movie with extensive underwater scenes. Connery is near his peak here and very comfortable in his role, and he does an excellent job.

There were things in *Tomorrow Never Dies* that I didn't care for, but I still thought it was worthy of ninth place in the list. Brosnan, as usual, was brilliant. The plot was a little over-the-top, and the motorcycle chase, while suspenseful, was a bit too long. News mogul Elliot Carver was sinister and not very likeable, and he played the part well. One of the highlights of the film was the HALO jump, but the BMW 750iL and the stealth ship were also quite amazing.

This brings us to the movie that started it all: *Dr. No*. Many of the elements that made the later Bond movies so successful were here, but there were no gadgets. It was more realistic than many of the subsequent films, but it still had a lot of action. I believe it was a first-rate adventure film.

My next selection might be a little controversial. Many people find *Licence to Kill* too serious and dark for a Bond movie, but I don't agree. True, it did have some gruesome parts, most of which involved sharks, but it had one of the best villains in the series, and one of the best Bond girls. Timothy Dalton was superb in the film. The truck chase at the end has to rate as one of the best action sequences in the Bond films. And it was delightful to see Q in an extended role.

For number 12 I have selected *GoldenEye*. It was Pierce Brosnan's first Bond movie, and it had a tremendous pretitle sequence: the bungee jump off the dam in Siberia. The plot was a little confused, and although it contained a lot of action, it seemed disjointed in places.

I'm sure I'm going to cause some controversy by placing *Die Another Day* so far down on the list. It had never-ending action throughout, but there may simply have been too much. It hardly gives you time to breathe. It had some great cars and amazing stunts in it, but I didn't like that so many scenes were stolen from earlier Bond movies. The first part of the film was the best.

Octopussy is a movie that grows on you. It had a tremendous pretitle sequence with the minijet *Acrostar*, but the first time I saw it I was a little disappointed. After seeing it several times, however, I have come to appreciate it, and now feel that it is one of Moore's better efforts. There is little slapstick in it and a lot of suspense in the ending.

We now come to *Moonraker*. It was a tremendous commercial success, and as a scientist I was looking forward to seeing it. As hoped, there was a lot of interesting science in it, even if some of it was wrong. I particularly enjoyed the finale in space, but like many other people I feel that the slapstick destroyed the movie. It could have been a great movie—one of the better ones—but it wasn't, and I was disappointed.

The next two on my list are both Sean Connery films, and as you will see in the next section I feel he was one of the best Bonds. He was getting a little old for the part in *Diamonds Are Forever*, however. Still, I thought it was a terrific movie; it just wasn't as good as some of his previous efforts. And again, there was nothing wrong with *You Only Live Twice*. It was an excellent adventure film with breathtaking scenery, and Sean Connery did a wonderful job in his role.

I won't say much about the final three movies, but I will say that I enjoyed all the Bond films but one has to come last in the list. I particularly liked the emphasis on solar energy in *The Man with the Golden Gun*. And who wouldn't be impressed with the stunt involving the car twisting off the broken bridge?

Best Bond Actors

1. Sean Connery
2. Pierce Brosnan
3. Roger Moore
4. Timothy Dalton
5. George Lazenby

Maybe it's because I saw Sean Connery first, and he seemed so perfect in the role that I picked him as number one. After Connery, there's no doubt that every other Bond was compared with him. He had everything going for him: he could be ruthless and cold, but he was also the master of one-line jokes that took the edge off some of his "darker" deeds. He was superbly self-confident—a man's man, but also the type that would attract women. And something else I felt was important: Connery had the best voice of all the Bonds.

Brosnan is a close second. He had many of the same positive features that Connery had, and with his looks some may have found him even more appealing. He was tough, but not overserious, and he could be humorous. And, he was believable. He was particularly good in the action scenes and he had a lot of them.

Moore is a bit of an enigma. He was in some of the very best Bond movies, but he was also in all three of the worst. There's no doubt that he was charming and that he played the role quite differently from Connery. Because of his humorous, rather light-hearted approach to his role, it was difficult to take him seriously at times. Still, he was in two of my favorite movies, and he did an excellent job in them.

Timothy Dalton has taken a beating from many Bond fans, mostly because he was so serious. With the slapstick of the last few Moore films, however, he was a welcome change for many. I thoroughly enjoyed both of his movies—*The Living Daylights*, in particular. While lacking some of the charisma and charm of Connery and Brosnan, he was a believable British agent, and perhaps closer to Fleming's Bond than any of the others.

Finally we have Lazenby. His lack of acting experience showed, but he has to be given credit for excellent action scenes. Furthermore, the one Bond film that he starred in (*On Her Majesty's Secret Service*) was an excellent movie despite his acting.

Best Villains

When it comes to choosing the best villains, there are several ways to pick them. We could ask who was the best actor, or who was the "scariest," in other words, who really sent chills up your back? As much as possible, I've tried to take all of these into consideration.

1. Robert Davi as Franz Sanchez in *Licence to Kill*
2. Robert Shaw as Red Grant in *From Russia with Love*
3. Gert Frobe as Auric Goldfinger in *Goldfinger*
4. Michael Lonsdale as Hugo Drax in *Moonraker*
5. Jonathan Pryce as Elliot Carver in *Tomorrow Never Dies*

Sanchez was loyal to anyone who was loyal to him, but watch out if you crossed him. He was quite believable as a villain and about as ruthless as you could get.

In Fleming's book *From Russia with Love*, the first forty pages are devoted to building up Grant as a ruthless, cold-hearted killer. After reading them you wonder how Bond is going to manage if he meets him. They come face to face on the Orient Express, and their fight scene is one of the best action sequences.

This brings us to Goldfinger. It wasn't that Goldfinger himself was so scary, but he had one of the best henchmen in the series, Oddjob. Bond was obviously no match for him physically, but in the end he managed to overcome him with cunning instead of strength. Goldfinger had one of the best lines in the series. When the laser was approaching his crotch Bond asked, "Do you expect me to talk?" "No, Mr. Bond," replied Goldfinger. "I expect you to die."

Michael Lonsdale's portrayal of Drax in *Moonraker* stands out as one of the few strong parts in the film. Drax had many of the best lines, although he wasn't as ruthless as some of the other villains.

Finally, we have Elliot Carver in *Tomorrow Never Dies*. Like Drax he had an obsession. He was an evil man who was possessed, and he played the role well.

Best Bond Girls

1. Carole Bouquet as Melina Havelock in *For Your Eyes Only*
2. Cary Powell as Pam Bouvier in *Licence to Kill*
3. Barbara Bach as Anya Amasova in *The Spy Who Loved Me*
4. Diana Rigg as Tracy Di Vicenzo in *On Her Majesty's Secret Service*
5. Claudine Auger as Domino Derval in *Thunderball*

If the Bond girls were judged only on glamour and sexiness I'd have to mention a few more. Ursula Andress as Honey Ryder was hard to beat in *Dr. No*. And who could forget Shirley Eaton as Jill Masterson in *Goldfinger*. Honor Blackman did an excellent job as Pussy Galore in the same movie. And for pure glamour it's hard to forget Jill St. John as Tiffany Case in *Diamonds Are Forever*, and for pure sexiness, Halle Berry in *Die Another Day*.

My first choice is no doubt influenced by the fact that *For Your Eyes Only* was one of my favorite Bond movies. Melina was set on revenge, and the part she played was believable. She had both beauty and brains, and she played a key role in the movie.

Another independent woman was Pam Bouvier in *Licence to Kill*. She had no intentions of playing second fiddle to Bond, and when he told her to leave and go back to the United States, she ignored him. And as it turned out, Bond needed her in the final scenes. She was a tough lady but also very cute.

Much of the same thing can be said of Barbara Bach in *The Spy Who Loved Me*. She was independent and a little headstrong. The problem in her case, however, was that she was not a good actress, and many of her lines seemed stiff.

In contrast, Diana Rigg of *On Her Majesty's Secret Service* was a seasoned actress with several years on the TV show *The Avengers*, and she played her part well. Finally, in fifth place I have Domino Derval of *Thunderball*. The relationship that develops between Bond and Domino in the movie is interesting and believable.

Best Cars

1. Aston Martin DB5 in *Goldfinger*
2. Lotus Esprit in *The Spy Who Loved Me*
3. BMW 750iL in *Tomorrow Never Dies*
4. Aston Martin V12 Vanquish in *Die Another Day*
5. BMW Z3 in *GoldenEye*

Ask most people about Bond's cars, and the one they're most likely to remember is the Aston Martin in *Goldfinger*. It was called the "most famous car in the world," and with good reason. It was so popular it was later used in *GoldenEye, Tomorrow Never Dies,* and in *Die Another Day*.

Moore's car was the Lotus Esprit, and the most famous appearance was as the "submarine car" in *The Spy Who Loved Me*. I'm sure it surprised everyone when the car disappeared into the ocean and became a fully operational submarine. It was one of the highlights of the film.

The BMW 750iL was a different car for Bond; he usually drove a sports car. One of its more interesting features was that it could be driven by remote control. And finally, the Z3 in *GoldenEye* was used mostly for publicity, but it definitely had eye appeal.

Best Gadgets

I'm not sure you can call the first two items in the list "gadgets," but they are two of the most intriguing conveyances in the series and they deserve some attention.

1. "Little Nellie" in *You Only Live Twice*
2. *Acrostar* in *Octopussy*
3. The jet pack in *Thunderball*
4. The bullet-deflecting watch in *Live and Let Die*
5. The X-ray glasses in *The World Is Not Enough*

I also have to give honorable mention to the key ring in *The Living Daylights*.

It's hard to beat "Little Nellie" and her arsenal of weapons. She was fast and highly maneuverable, and the action scene involving her was a highlight. In the same vein we have the *Acrostar*, which provided an excellent pre-title sequence in *Octopussy*.

The jet pack in *Thunderball* came as quite a surprise and generated a lot of interest in the devices for the next few years. As for the bullet-deflecting watch, I'm surprised that Bond didn't use it more.

Best Chase Scenes

1. The Aston Martin chase in *Goldfinger*
2. Race for the border in *The Living Daylights*
3. The tanker truck chase in *Licence to Kill*
4. The ice chase in *Die Another Day*
5. The motorcycle chase in *Tomorrow Never Dies*

I already talked about car chases in considerable detail in chapter 7, so I won't say any more about them. I should, however, give honorable mention to two other chases: the Las Vegas chase in *Diamonds Are Forever* and the downhill chase in *GoldenEye*.

Best Stunts

1. The sky jump in *Moonraker*
2. The bungee jump in *GoldenEye*
3. Bond and Necros hanging from netting at the back of a plane in *The Living Daylights*
4. Skiing off the cliff in *The Spy Who Loved Me*
5. Bond going over a cliff on a motorcycle and catching a plane in *GoldenEye*

Honorable mention also has to go to Bond hanging on to a plane in *Octopussy*.

Again, I've talked about the stunts in considerable detail in earlier chapters. There are so many excellent ones it was difficult to place them in order of merit, but the sky jump in *Moonraker* was so spectacular it had to be placed first.

Best Action Scenes

1. Fight with Red Grant in *From Russia with Love*
2. *Thunderball* underwater battle
3. Fight at the end of *The Spy Who Loved Me*
4. Fight at the end of *You Only Live Twice*
5. Ski chase in *On Her Majesty's Secret Service*

So this brings us to the end of our lists, and also to the end of the book. As I look back over it I find there is still a lot I would like to say. Most of all, though, I would like to say that the Bond films were a delight, but I'm sure, as a Bond fan, I don't need to tell you that. Let's hope they continue for a long time to come.

Bibliography

Asimov, Isaac. *The History of Physics.* New York: Walker and Co., 1966.

Berman, Arthur. *The Physical Principles of Astronautics.* New York: Wiley, 1961.

Black, Michael. *Bungy Jumping.* May 2001. Available at: www.extremz.com.

Brosnan, John. *James Bond in the Cinema.* London: Tantivy Press, 1981.

Chapman, James. *Licence to Thrill.* New York: Columbia University Press, 2000.

Cork, John, and Bruce Scivally. *James Bond: The Legacy.* New York: Harry Abrams, 2002.

Di Leo, Michael. *The Spy Who Thrilled Us.* New York: Limelight Editions, 2002.

Dougall, Alastair, and Roger Stewart. *James Bond: The Secret World of 007.* New York: DK Publishing, 2000.

Fleming, Ian. *Diamonds Are Forever.* New York: Macmillan, 1956.

——. *Dr. No.* New York: Macmillan, 1958.

——. *From Russia with Love.* New York: New American Library, 1957.

——. *Goldfinger.* New York: New American Library, 1959.

——. *Moonraker.* New York: Macmillan, 1955.

——. *On Her Majesty's Secret Service.* New York: New American Library, 1963.

——. *Thunderball.* New York: New American Library, 1961.

——. *The Man with the Golden Gun.* New York: New American Library, 1965.

——. *The Spy Who Loved Me.* New York: Penguin, 2002.

——. *You Only Live Twice.* New York: New American Library, 2002.

Halmark, Clayton. *Lasers: The Light Fantastic*. Blue Ridge: Tab Books, 1979.

Heavens, O. S. *Lasers*. New York: Scribners, 1971.

Heckman, Philip. *The Magic of Holography*. New York: Atheneum, 1986.

Lind, David, and Scott Sanders. *The Physics of Skiing*. New York: Springer-Verlag, 1996.

Parker, Barry. *Einstein's Vision*. Amherst, MA: Prometheus, 2004.

———. *The Isaac Newton School of Driving: Physics & Your Car*. Baltimore: Johns Hopkins University Press, 2003.

Pearson, John. *The Life of Ian Fleming*. New York: McGraw-Hill, 1966.

Pfeiffer, Lee, and Dave Worrall, *The Essential Bond*. New York: HarperCollins, 2000.

Rhodes, Richard. *The Making of the Atomic Bomb*. New York: Simon and Schuster, 1986.

Rubin, Steven Jay. *The Complete James Bond Encyclopedia*. Chicago: Contemporary Books, 2003.

Smelling, O. F. *007 James Bond: A Report*. New York: Signet Books, 1964.

Internet Sites

www.avuhub.net

www.biowaves.com

www.blackmagic.com

www.brook.edu

www.carenthusiast.com

www.comfuture.com

www.cord.edu

www.extremz.com

www.fas.org

www.geocities.com

www.howstuffworks.com

www.hypertextbook.com

www.istp.gfsc.nasa.gov

www.jetsprint.org

www.madsci.org

www.math.utah.edu

www.mindspring.com

www.rifleshootermag.com

www.space.edu

www.spybusters.com

www.swssec.com

www.universalexports.net/movies

Index